The Art and Science of Construction Claims

SHREEDHAR BHAT FRICS|FICCP|FIIV

The views expressed in this book are based on professional experience and practical interpretation of construction project realities. The contents are intended for educational and professional discussion purposes and should not be construed as legal advice.

ISBN:

DEDICATION

To all project professionals who continue to balance contractual obligations with execution realities under increasingly complex project environments.

Also dedicated to the pursuit of fairness, professionalism, and practical wisdom in construction projects.

CONTENTS

ACKNOWLEDGMENTS

This work is the culmination of decades of professional exposure across construction projects, commercial management, claims administration, and dispute resolution.

The author expresses sincere gratitude to the many professionals, colleagues, mentors, engineers, commercial teams, planners, contractors, consultants, and project personnel whose practical experiences and interactions contributed, directly and indirectly, to the insights reflected in this manuscript.

Construction projects are shaped not only by drawings, contracts, and schedules, but by people, decisions, constraints, and evolving realities. Many of the reflections presented in this book emerged from observing how projects behave in practice — often very differently from how they are intended to function contractually.

The author also acknowledges the contribution of the broader professional community involved in:

- construction claims;
- project controls;
- commercial management;
- arbitration;
- and infrastructure delivery.

Their continuing engagement with the practical challenges of modern projects remains an important source of collective industry learning.

Special appreciation is extended to professionals and peers who encouraged the development of this manuscript and recognised the importance of documenting practical execution realities alongside contractual theory.

Finally, the author acknowledges that every project — whether successful or troubled — contributes valuable lessons. This manuscript is ultimately an attempt to capture some of those lessons in a structured and reflective form for the benefit of the wider construction industry.

Preface

Construction claims are often discussed through contractual clauses, legal principles, and delay methodologies. Yet, anyone who has spent substantial time within live projects understands that claims rarely arise from contractual language alone.

Projects are shaped by:

- incomplete information;
- evolving site conditions;
- environmental realities;
- logistical constraints;
- financial pressures;
- behavioural decisions;
- and human responses to uncertainty.

Over the course of professional practice, it became increasingly evident that many claim situations cannot be fully understood merely through contractual interpretation. In many cases, the real issues emerge from the interaction between:

- what was assumed during tender;
- what became possible during execution;
- and how parties responded commercially and procedurally when conditions changed.

This manuscript was therefore developed not as a purely legal or academic commentary, but as a reflective and practice-led examination of how construction claims evolve in real projects.

The intention is not to advocate adversarial claim behaviour, but rather to encourage:

- balanced risk understanding;
- practical interpretation;
- better documentation practices;
- and commercially mature decision-making.

The case studies and discussions presented throughout the book are intended to illustrate broader industry patterns rather than isolated project events.

Several recurring themes emerge repeatedly across projects:

- disruption is often cumulative rather than singular;
- productivity loss is frequently real but difficult to quantify;

- contractual clauses are sometimes interpreted without adequate consideration of practical feasibility;
- and many otherwise valid claims fail due to procedural or strategic shortcomings.

An equally important observation is that projects are increasingly influenced by behavioural and organisational factors. The quality of:

- data systems;
- internal coordination;
- decision timing;
- and commercial strategy

often influences claim outcomes as much as entitlement itself.

This book therefore attempts to bridge the gap between:

- contractual theory;
- project execution realities;
- and commercial behaviour.

The discussions contained in this manuscript are based substantially on practical experience developed across residential, commercial, industrial, and infrastructure projects over several decades.

If the work contributes toward:

- more balanced claim assessment;
- improved project administration;
- and deeper professional reflection within the industry,

then it will have achieved its intended purpose.

Executive Summary

Construction claims are often perceived primarily as contractual or legal disputes centred on entitlement, delay, and compensation. In practice, however, the reality of construction claims is significantly more complex. Project outcomes are shaped not only by contractual provisions, but also by execution realities, behavioural responses, commercial pressures, environmental conditions, organisational capability, and the timing of decisions.

This book examines construction claims through a practical, experience-driven, and case-oriented perspective, focusing particularly on the distinction between theoretical risk allocation and actual project execution.

A central theme of the book is that disruption in construction projects rarely arises from a single isolated event. Instead, it often develops gradually through cumulative operational inefficiencies such as:

- fragmented work fronts;
- approval delays;
- restricted access;
- weather-related recovery efforts;
- logistics and supply-chain interruptions;
- labour instability;
- and financial stress arising from delayed payments.

While many of these impacts may not immediately appear significant in isolation, their cumulative effect can substantially reduce productivity, increase execution cost, and alter project behaviour.

The book therefore distinguishes clearly between:

- **delay**, which affects completion timelines; and
- **disruption**, which affects productivity and efficiency, often without immediately impacting completion dates.

This distinction is critical because many projects experience substantial disruption even where formal Extension of Time (EOT) entitlement is limited or disputed.

The study further examines how contractual mechanisms such as:

- “fit for purpose” obligations;
- “duty of care” provisions;
- “indicative data” disclaimers;

- and strict notice requirements

are frequently interpreted in a manner that transfers broad execution risk to Contractors, sometimes without sufficient consideration of practical feasibility, execution constraints, or the adequacy of information available at tender stage.

Through multiple real-world case illustrations, the book demonstrates that:

- the success or failure of claims is often determined less by pure entitlement and more by:
    - timing of notices;
    - quality of documentation;
    - internal organisational capability;
    - commercial negotiation strategy;
    - and the ability to demonstrate causation and productivity impact.

The case studies highlight recurring industry patterns, including:

- mismatch between tender assumptions and actual site conditions;
- misuse or overextension of contractual clauses;
- undervaluation of time-related benefits in Value Engineering;
- failure to recognise indirect disruption impacts;
- and procedural failures leading to loss of otherwise valid entitlements.

A further recurring theme throughout the book is the gap between:

- **entitlement in principle**, and
- **recoverability in practice**.

Many claims that may appear valid from a contractual perspective are either:

- never pursued;
- commercially compromised;
- procedurally barred;
- or insufficiently substantiated.

The book also explores the behavioural and commercial dimensions of construction claims, including:

- reluctance toward open-book cost disclosure;
- organisational misalignment between project teams and management;
- delayed decision-making;
- and the influence of cash-flow pressures on project productivity and claim behaviour.

In addition to analysing claim failures, the book presents a structured practical framework for developing disruption claims through:

1. identification of triggering events;
2. demonstration of changed working conditions;
3. evidence of productivity loss;
4. establishment of causation;
5. and quantification of financial impact.

The study ultimately argues that construction claims cannot be effectively understood through contractual interpretation alone. Successful claim management requires integration of:

- technical understanding;
- operational realities;
- commercial judgement;
- behavioural awareness;
- and structured data-driven analysis.

The book concludes by examining the future direction of construction claims in an increasingly digital and data-driven industry environment. It highlights the growing importance of:

- real-time project data;
- digital documentation systems;
- integrated project controls;
- and proactive claim management approaches.

At its core, this book advances a broader professional insight:

Construction claims are not merely legal disputes arising from contracts; they are reflections of how projects behave under uncertainty, disruption, commercial pressure, and human decision-making.

Accordingly, the work seeks to bridge the gap between:

- contractual theory;
- project execution;
- and commercial reality,

while providing practitioners, consultants, contractors, employers, arbitrators, and project professionals with a more grounded and holistic understanding of construction claims in practice.

Framework Overview

Conceptual Structure of the Book

Construction claims are often approached as isolated contractual events requiring legal interpretation and procedural compliance. However, practical project experience demonstrates that claims evolve through the interaction of multiple interconnected factors, including:

- execution realities;
- environmental and logistical constraints;
- behavioural responses;
- commercial pressures;
- organisational capability;
- and contractual administration practices.

Accordingly, this book is structured not merely as a technical claims manual, but as an integrated framework examining how construction claims develop, evolve, succeed, fail, and ultimately influence project outcomes.

The chapters are arranged progressively to move from:

- conceptual foundations,

  to:

- operational disruption mechanisms,

  then toward:

- practical case analysis,
- claim-development methodologies,
- and broader industry implications.

The framework of the book can therefore be understood through five interconnected layers.

Layer 1 — Foundational Understanding (Chapters 1–2)

The opening chapters establish the intellectual foundation of the book by examining:

- the evolving nature of construction claims;
- the distinction between delay and disruption;
- and the limitations of analysing claims purely through contractual entitlement.

These chapters introduce a central premise of the book:

Construction claims are not governed solely by contract clauses, but by the interaction between contractual provisions and practical execution realities.

A clear distinction is developed between:

- **delay**, which affects project completion; and
- **disruption**, which affects productivity and efficiency.

This distinction becomes fundamental to all subsequent analysis.

The opening framework also introduces:

- the concept of hidden disruption;
- cumulative inefficiencies;
- and the gap between entitlement and practical recoverability.

Layer 2 — Hidden Disruption and Contractual Limits (Chapters 3–4)

These chapters examine the less-visible operational and contractual factors that frequently trigger construction claims.

The analysis demonstrates that disruption often arises not from dramatic singular events, but from:

- gradual productivity deterioration;
- fragmented execution;
- weather recovery impacts;
- logistical restrictions;
- supply-chain interruptions;
- approval delays;
- financial pressures;
- and practical execution constraints.

Chapter 3 focuses on:

- hidden disruption triggers,
- environmental realities,
- logistical inefficiencies,
- and cumulative operational impacts.

Particular emphasis is placed on:

- post-event recovery;
- continuity of working conditions;
- and productivity deterioration that is often overlooked in conventional claim assessment.

Chapter 4 examines:

- “fit for purpose” obligations;
- “duty of care”;

- and limitations of broad contractual risk-transfer mechanisms.

The chapter argues that:

contractual clauses cannot be interpreted independently of practical execution feasibility and reasonable foreseeability.

Together, these chapters establish the operational and contractual environment within which construction claims actually emerge.

Layer 3 — Commercial and Behavioural Realities (Chapters 5–6)

The next layer of the framework shifts focus from contractual theory toward:

- commercial behaviour;
- organisational dynamics;
- and practical project decision-making.

Chapter 5 explores how claims are shaped by:

- working-capital pressures;
- reluctance toward open-book disclosure;
- negotiation behaviour;
- cost of dispute resolution;
- organisational limitations;
- and delayed decision-making.

The analysis highlights a recurring industry reality:

Many claims fail not due to absence of entitlement, but due to behavioural, procedural, and commercial limitations.

The chapter further examines:

- deferred EOT determination practices;
- internal organisational misalignment;
- and the growing importance of data culture and digital systems.

Chapter 6 forms the experiential core of the book through detailed case studies drawn from real project scenarios.

These case illustrations collectively demonstrate:

- how disruption develops in practice;
- how claims are often mischaracterised;
- how procedural failures extinguish entitlement;
- and how commercial strategy frequently determines outcomes more than contractual

interpretation alone.

The case studies also reinforce recurring themes such as:

- tender-versus-execution mismatch;
- hidden productivity loss;
- time-versus-cost misalignment;
- and behavioural influences on claim recovery.

Layer 4 — Practical Claim Development Framework (Chapters 7–8)

Having established the operational and behavioural context of claims, the book then transitions toward practical application.

Chapter 7 presents a structured framework for disruption claims based on:

1. identification of triggering events;
2. demonstration of changed working conditions;
3. evidence of productivity loss;
4. establishment of causation;
5. quantification of financial impact.

The framework bridges the gap between:

- project realities;
- and formal claim substantiation.

Particular emphasis is placed on:

- causation logic;
- productivity analysis;
- and differentiation between delay and disruption impacts.

Chapter 8 distils the broader governing principles emerging from the earlier chapters and case studies.

These principles include:

- reasonable versus unlimited risk transfer;
- practical limits of contractual clauses;
- behavioural dimensions of claim recovery;
- importance of documentation;
- and the distinction between entitlement and recoverability.

This section serves as the conceptual synthesis of the book.

Layer 5 — Future Direction and Industry Evolution (Chapters 9–10)

The final layer examines the future evolution of construction claims in an increasingly:

- digital;
- data-driven;
- and integrated project environment.

Chapter 9 explores:

- digital project controls;
- BIM and integrated systems;
- real-time productivity monitoring;
- and the growing role of structured project data in claim management and dispute resolution.

The chapter argues that:

future claims management will increasingly depend on data integrity, transparency, and proactive project controls rather than retrospective reconstruction alone.

Chapter 10 concludes the book by integrating:

- contractual analysis;
- execution behaviour;
- commercial realities;
- and organisational dynamics

into a broader professional understanding of construction claims.

The concluding insight of the book is that:

Construction claims are ultimately reflections of how projects behave under uncertainty, disruption, commercial pressure, and human decision-making.

Integrated Conceptual Flow of the Book

The overall progression of the book may therefore be understood as follows:

Layer	Focus	Chapters
Foundational Concepts	Delay, disruption, entitlement	1–2
Hidden Disruption & Contractual Limits	Operational realities and risk allocation	3–4
Commercial & Behavioural Dynamics	Organisational and practical claim behaviour	5–6
Practical Claim Framework	Claim development and governing	7–8

Layer	Focus	Chapters
	principles	
Future Direction	Data-driven claims and industry evolution	9–10

Central Theme of the Book

Across all chapters, one consistent theme emerges:

Construction claims cannot be fully understood through contractual interpretation alone.

They must also be understood through:

- execution realities;
- productivity behaviour;
- commercial pressures;
- organisational capability;
- and the practical limitations under which projects operate.

The framework presented in this book therefore seeks to bridge:

- contractual theory;
- construction execution;
- and commercial reality,

to provide a more practical and integrated understanding of construction claims in modern projects.

Chapter 1

Introduction — Construction Claims Beyond Contractual Theory

1.1 The Evolving Nature of Construction Claims

Construction claims have traditionally been analysed through contractual interpretation. Conventional discussions usually focus upon clauses, compensable events, contractual remedies, notice provisions, and legal entitlement.

However, practical project execution demonstrates that claim outcomes are rarely determined by contractual wording alone.

Construction projects operate within highly dynamic environments shaped by:

- engineering realities;
- environmental conditions;
- labour behaviour;
- procurement constraints;
- logistical limitations;
- financial pressures;
- organisational culture;
- and timing of decisions.

As projects become larger, more complex, and increasingly interconnected, the gap between contractual assumptions and execution realities becomes more visible.

Modern construction contracts attempt to allocate substantial execution risk to Contractors through mechanisms such as:

- "fit for purpose" obligations;
- "duty of care" provisions;
- indicative site-data disclaimers;
- strict notice and time-bar requirements;
- and broad transfer of execution responsibility.

From a contractual perspective, such provisions may appear commercially logical. In practice, however, their application often extends beyond what can reasonably be foreseen, verified, quantified, or commercially priced at the tender stage.

A Contractor may reasonably accept responsibility for execution, but cannot always realistically anticipate:

- incomplete or non-representative geotechnical information;
- cumulative productivity loss arising from fragmented work fronts;
- prolonged recovery efforts following adverse weather;
- indirect inefficiencies caused by delayed approvals;
- or labour instability arising from delayed payments.

Consequently, many disputes arise not because of deliberate contractual non-compliance, but because the actual conditions under which the work is executed differ materially from the assumptions upon which pricing, planning, and methodology were originally based.

Construction claims therefore cannot be fully understood through legal analysis alone.

They must also be understood through:

- execution behaviour;
- operational feasibility;
- commercial realities;
- organisational capability;
- and practical project dynamics.

1.2 From Contract Administration to Execution Reality

A recurring issue in construction projects is the disconnect between contract administration and actual site behaviour.

Contracts generally assume that:

- work fronts will remain reasonably available;
- approvals will be issued within planned timelines;
- access conditions will remain manageable;
- and baseline assumptions will broadly remain valid.

In practice, execution rarely progresses in such a linear or controlled manner.

Minor disruptions often accumulate gradually through:

- resequencing of activities;
- restricted access;
- repeated mobilisation and demobilisation;
- fragmented execution;
- environmental interference;

- or operational inefficiencies.

Individually, these events may appear manageable. Collectively, however, they can significantly affect productivity, continuity, and cost.

Importantly, many such impacts remain commercially underestimated because they do not always manifest as obvious delay to project completion.

The Contractor may continue progressing the works through:

- acceleration;
- deployment of additional labour;
- overtime;
- parallel execution;
- or inefficient utilisation of resources.

The project may therefore appear to continue progressing while hidden productivity loss accumulates beneath the surface.

This distinction between visible delay and hidden disruption forms one of the central themes of this book.

1.3 Behavioural and Commercial Dimensions of Claims

Construction claims are frequently perceived as technical or legal exercises. In practice, however, behavioural and commercial factors often play an equally decisive role.

Many valid claims are:

- never pursued;
- poorly framed;
- commercially compromised;
- or procedurally weakened.

This commonly occurs due to:

- reluctance to disclose internal cost structures;
- weak contemporaneous documentation;
- organisational disconnect between site and head office;
- delayed decision-making;
- fear of damaging commercial relationships;
- and the practical cost of dispute resolution.

As a result, entitlement in principle does not always translate into recovery in practice.

This gap between entitlement and recoverability is one of the most important practical realities in construction claim management.

1.4 Objectives of This Book

This book therefore adopts a practice-oriented and experience-based approach to construction claims.

Rather than analysing claims purely through contractual doctrine, the discussion focuses on:

- how disruption actually develops within projects;
- how productivity loss manifests operationally;
- how claims are framed or misframed;
- why valid claims frequently fail;
- and how behavioural and commercial realities influence outcomes.

The objective is not merely to explain contractual entitlement, but to bridge the gap between contractual theory and execution reality.

Chapter 2

Delay vs Disruption — A Fundamental Distinction

A fundamental difficulty in construction claims practice arises from the tendency to treat delay and disruption as interchangeable concepts.

Although closely related, the two represent materially different forms of project impact.

Delay is primarily concerned with time.

Disruption is primarily concerned with efficiency.

This distinction is more than academic. It directly affects:

- entitlement;
- methods of analysis;
- evidentiary requirements;
- and commercial recovery.

In practice, projects frequently suffer substantial disruption without corresponding delay to completion. Contractors may accelerate activities, deploy additional resources, extend working hours, or fragment work sequences in order to maintain milestone commitments.

From an external perspective, the project may appear to remain on schedule.

Operationally, however, the work may already be suffering severe productivity deterioration.

The same resources may produce substantially lower output under disrupted conditions.

Labour movement becomes inefficient.

Material handling increases.

Continuous work fronts become fragmented.

Activities originally intended to proceed sequentially begin competing simultaneously for limited access, supervision, logistics, and resources.

Under such circumstances, the project continues, but no longer progresses efficiently.

This is the operational essence of disruption.

Conversely, delay becomes visible when the impact extends into activities affecting the critical path and ultimately influences the project completion date.

While delay is generally easier to identify through programme analysis, disruption is often more difficult to establish because its effects are dispersed across:

- productivity;
- resource utilisation;
- sequencing;
- and operational continuity.

For this reason, disruption claims frequently become more complex than delay claims, even where the financial impact is substantial.

Understanding this distinction is essential because many disputes arise not from absence of impact, but from mischaracterisation of the nature of that impact. To understand the distinction between Delay & Disruption it is important to understand the nature, relationship and consequences of misunderstanding which is explained below.

2.1 Nature of Delay

Delay primarily concerns impact on project completion.

A delay exists where progress of work is affected in such a manner that:

- the contractual completion date;
- milestone dates;
- or critical path activities; are impacted.

Delay analysis therefore focuses principally on:

- time;
- sequencing;
- and programme logic.

The essential question in delay analysis is:

Did the event affect the planned completion of the project?

Accordingly, delay claims are commonly evaluated through programme-based techniques such as:

- Critical Path Method (CPM) analysis;
- time impact analysis;

- window analysis;
- or as-planned versus as-built comparisons.

Delay itself may arise from multiple sources, including:

- delayed approvals;
- late drawings;
- restricted access;
- variation instructions;
- environmental conditions;
- labour disruptions;
- or supply-chain interruptions.

However, irrespective of the cause, delay analysis fundamentally examines:

- impact on planned completion.

Where delay affects the critical path, the Contractor may become entitled to:

- Extension of Time (EOT);
- and potentially prolongation-related compensation.

2.2 Nature of Disruption

Disruption differs fundamentally from delay.

Disruption concerns deterioration in productivity and efficiency during execution.

A project may continue progressing and even achieve completion within time, while still suffering severe disruption.

Disruption occurs where work is executed:

- inefficiently;
- under altered conditions;
- or through fragmented and unplanned methodologies.

It manifests operationally through:

- reduced labour output;
- repeated mobilisation and demobilisation;
- inefficient sequencing;
- interruption of workflow continuity;
- increased supervision requirements;
- excessive material handling;

- and deterioration in resource productivity.

The essential question in disruption analysis is therefore:

Was the Contractor forced to execute the work less efficiently than originally planned?

A Contractor may:

- accelerate work;
- deploy additional resources;
- or absorb inefficiency; in order to avoid overall project delay.

In such situations:

- delay to completion may not occur; while:
- substantial disruption cost may still arise.

Disruption is therefore fundamentally linked to:

- productivity loss; rather than:
- completion impact alone.

2.3 Relationship Between Delay and Disruption

Although delay and disruption are distinct concepts, they frequently interact within construction projects.

Disruption can eventually lead to delay where productivity deterioration affects critical-path activities.

Similarly, delay events often create secondary disruption impacts through:

- fragmented execution;
- resequencing;
- inefficient work fronts;
- and repeated resource deployment.

Accordingly, a single event may generate both:

- time-related consequences;
- and productivity-related consequences.

For example:

- delayed access to work areas may delay completion; while simultaneously:
- forcing labour and equipment to work inefficiently across fragmented fronts.

In such cases, the Contractor may potentially be entitled to:

- Extension of Time for delay impact;

- and additional compensation for disruption-related inefficiency.

However, these impacts must still be analysed separately.

Failure to distinguish between:

- completion delay; and:
- productivity deterioration; frequently results in confusion and weak claim presentation.

2.4 Why Disruption Claims Are More Difficult

In practice, disruption claims are generally more difficult to establish than delay claims.

This difficulty arises because:

- delay is often visible through programme movement; whereas:
- disruption is dispersed across operational performance.

A delayed activity can often be identified relatively clearly through:

- programme updates;
- milestone slippage;
- or critical-path movement.

Disruption, however, may manifest gradually through:

- reduced output;
- increased labour effort;
- intermittent stoppages;
- reduced equipment utilisation;
- or fragmented workflow.

Such impacts are operationally real but often difficult to isolate precisely.

Additionally, disruption analysis frequently requires:

- productivity measurement;
- comparison of unaffected and affected periods;
- and elimination of alternative explanations.

This creates significant evidentiary complexity.

Employers and Consultants often argue that:

- reduced productivity resulted from Contractor inefficiency;
- poor planning;
- or inadequate resource deployment.

Accordingly, successful disruption claims require:

- stronger causation analysis;
- more detailed records;
- and clearer linkage between event and productivity loss.

2.5 Practical Consequences of Misunderstanding the Distinction

Failure to distinguish properly between delay and disruption frequently produces serious practical consequences.

These include:

- incorrect claim framing;
- inadequate substantiation;
- double counting of impacts;
- inconsistent quantification;
- and weakened negotiation positions.

For example, Contractors sometimes attempt to pursue disruption indirectly through:

- prolongation claims;
- inflated quantity claims;
- or general commercial assertions; rather than specifically demonstrating:
- productivity deterioration;
- and changed working conditions.

Similarly, Employers may incorrectly assume that:

- absence of delay automatically eliminates entitlement.

In reality:

- disruption may remain substantial even where completion is maintained through acceleration or additional resources.

Understanding the distinction between delay and disruption is therefore essential not only for technical analysis, but also for:

- effective claim strategy;
- commercial negotiation;
- and fair dispute resolution.

2.6 Conclusion

Delay and disruption represent distinct but interconnected dimensions of construction project impact.

Delay primarily concerns:

- impact on completion;
- programme movement;
- and critical-path activities.

Disruption primarily concerns:

- deterioration in productivity;
- loss of execution efficiency;
- and interference with planned working methods.

Although both frequently arise from the same events, they require separate evaluation because their operational and commercial consequences differ materially.

Modern construction projects increasingly demonstrate that substantial disruption may exist even where overall completion is maintained through acceleration, additional resources, or resequencing.

Accordingly, effective claims management requires clear identification of:

- time-related impact;
- productivity-related impact;
- and the relationship between the two.

Only through this distinction can:

- causation be properly analysed;
- entitlement be accurately framed;
- and claims be evaluated in a fair and technically defensible manner.

Chapter 3

Hidden Disruption Triggers in Construction Projects

In modern construction projects, disruption rarely arises from a single isolated event. More often, it develops gradually through the interaction of operational constraints, contractual ambiguities, environmental conditions, logistical limitations, and commercial pressures that interfere with planned execution.

Many of these influences are not expressly recognised within contractual frameworks. Certain conditions are treated as normal execution risks, while others remain concealed within broad contractual assumptions or general obligations. Yet, from an operational perspective, their cumulative effect on productivity, sequencing, labour efficiency, and continuity of work can be substantial.

A significant practical difficulty in construction claims is that disruption often evolves progressively rather than dramatically. Unlike an immediately identifiable delay event, disruption may emerge through small but continuous inefficiencies that gradually erode productivity across multiple work fronts.

Accordingly, understanding disruption requires a shift in perspective — from viewing projects solely through contractual milestones and completion dates to examining how work is actually executed under changing site realities.

The following sections examine some of the most common yet frequently underestimated disruption triggers encountered in practice.

3.1 Weather Impact and Post-Event Recovery

Weather-related disruption is often misunderstood because assessment tends to focus primarily on the duration or intensity of rainfall itself. In practice, however, the actual impact of weather extends far beyond the rainfall event.

Continuous drizzle, intermittent showers, and prolonged wet conditions frequently create greater operational inefficiency than short-duration extreme rainfall events. The disruption arises not merely from precipitation, but from the manner in which weather interferes with mobility, safety, material handling, continuity of operations, and post-event recovery.

On active construction sites, particularly during earthwork and structural phases, wet conditions gradually transform working surfaces into unstable and slippery zones. Labour movement becomes slower and more cautious. Equipment mobility reduces. Material handling cycles become fragmented. Activities that normally proceed continuously begin operating intermittently.

The impact becomes especially significant in earthwork operations. Excavated soil exposed to rainfall often becomes slushy and unstable, reducing excavation efficiency and slowing disposal operations. Excavators experience restricted movement, while trucks transporting excavated material face delays caused by deteriorated haul roads, waterlogging, and increased cycle times.

The disruption therefore affects not only excavation itself, but the entire material movement cycle — from excavation to transportation, disposal, and site restoration.

Similarly, reinforcement activities become increasingly inefficient during prolonged wet conditions. Wet steel bars reduce grip efficiency and create unsafe handling conditions. Labour productivity decreases not because work stops entirely, but because operations can no longer proceed with the continuity and rhythm originally planned.

In many cases, the most severe impact occurs not during rainfall, but after it.

This aspect is frequently underestimated during contractual evaluation and claim assessment.

Following rainfall, particularly in basements, retaining structures, partially completed slabs, and confined work areas, large volumes of slush, debris, and contaminated water accumulate within active work fronts. Floodwater may recede relatively quickly, yet the recovery process often continues for several days.

Labour teams are then diverted toward:

- cleaning reinforcement and formwork;
- removing accumulated slush;

- restoring access routes;
- re-establishing safe working surfaces;
- and recovering partially completed activities.

These efforts consume significant time and resources, but do not directly contribute to measurable physical progress. As a result, the actual disruption frequently remains under-recognised because conventional project measurement systems tend to capture completed quantities rather than recovery effort.

The practical consequence is that weather disruption often manifests as:

- fragmented execution;
- reduced productivity;
- intermittent stoppages;
- resequencing of activities;
- and prolonged recovery inefficiency.

This distinction becomes particularly important when evaluating entitlement under contracts that focus narrowly on rainfall thresholds or extreme weather definitions.

In practice, the true operational impact is governed not merely by the magnitude of rainfall, but by:

- duration of interference;
- continuity of adverse conditions;
- nature of ongoing activities;
- and the extent of post-event recovery required.

For example, prolonged wet conditions during foundation-stage earthwork may create manageable disruption due to limited work fronts. However, the same conditions during peak superstructure execution can severely affect multiple interconnected activities simultaneously.

Similarly, sustained wind conditions during foundation works may create limited operational impact, whereas identical wind conditions during bridge deck or elevated steel works may make execution unsafe and practically unworkable.

Accordingly, disruption assessment cannot be divorced from the stage of construction and the nature of activities being executed at that particular time.

Practical Insight — Misleading Severity

This can be understood through a simple practical observation:

A short-duration heavy downpour may appear severe but may cause limited disruption. In contrast, continuous drizzle throughout the day can significantly disturb activities due to sustained interference with working conditions.

In construction projects, disruption behaves in a similar manner.

Not all high-intensity events produce major operational impact. In many situations, moderate but sustained interference creates greater cumulative productivity loss because the work never fully stabilises. Labour movement slows continuously, equipment cycles become inefficient, and work fronts lose continuity.

The same principle applies beyond weather conditions.

A disruption occurring during the early stages of a project may have relatively limited impact because only a few work fronts are active. Likewise, during the final stages of completion, the number of concurrent activities may already have reduced.

However, during peak execution periods — when multiple interconnected work fronts operate simultaneously — even a seemingly minor disruption can create widespread cascading impact.

For instance, delay in the delivery of a critical glass sliding door for a project's main entrance may appear insignificant in isolation. Yet, if the project opening date has already been publicly committed, even a short disruption caused by transport restrictions or local access constraints can affect testing, finishing, commissioning, handover sequencing, and occupancy readiness.

Similarly, in coastal infrastructure projects, moderate but sustained wind conditions may not exceed contractual extreme-weather thresholds, yet may still render exposed working conditions unsafe for labour.

Where activities involve elevated bridge decks or superstructure works, prolonged moderate wind can create greater practical disruption than short-duration extreme events because execution continuity becomes impossible to maintain safely.

The critical factor is therefore not merely severity, but the relationship between:

- duration;
- continuity;
- stage of construction;
- work-front exposure;
- and operational interdependence.

Misjudging this relationship often leads to:

- incorrect claim framing;
- weak causation analysis;
- and underestimation of cumulative inefficiency.

The practical lesson is clear:

In construction disruption, duration and continuity frequently matter more than isolated intensity.

This also explains why weather-related disruption claims cannot be evaluated solely through meteorological data. Rainfall records may establish occurrence, but they rarely capture recovery effort, reduced workability, or loss of execution continuity.

Meaningful assessment therefore requires consideration not only of weather data, but also:

- site conditions;
- stage of execution;
- safety restrictions;
- operational continuity;
- and recovery effort.

Only then can the true extent of disruption be properly evaluated.

3.2 Supply Chain and Seasonal Availability Constraints

Environmental disruption is often closely linked to another underestimated vulnerability — time-sensitive availability of materials and resources.

Construction supply chains are rarely as flexible as contractual assumptions suggest. Many materials remain heavily dependent on seasonal accessibility, industrial production cycles, transportation continuity, and regional working patterns.

Consequently, even relatively minor delays in approvals or decision-making can trigger disproportionately large downstream disruption when they interfere with planned procurement windows.

Several construction materials remain inherently linked to environmental and seasonal conditions. Brick manufacturing depends on kiln operations and drying conditions. Quarry-based materials such as laterite stone become difficult to access during heavy rains. River sand extraction may be restricted seasonally due to environmental regulation or flooding conditions. Borrow earth and aggregate movement can become severely constrained by waterlogged access routes.

In addition to local environmental cycles, modern construction projects are increasingly influenced by global industrial and logistical calendars. Factory shutdowns during Chinese New Year, reduced industrial activity during Western summer holidays, transport congestion during festive periods, and labour migration during regional festivals all influence procurement continuity.

Experienced contractors typically plan around these known cycles.

The disruption arises when Employer-driven delays interfere with these carefully aligned procurement windows.

Under such conditions, the consequence is rarely limited to simple delay in delivery. Instead, projects experience:

- fragmented sequencing;
- intermittent work continuity;
- idle labour and equipment;
- repeated mobilisation and demobilisation;
- and cascading productivity loss across interconnected activities.

The resulting disruption therefore extends far beyond material delay itself.

In many projects, a missed procurement window gradually transforms into a broader execution inefficiency affecting multiple downstream activities and commercial outcomes.

This is particularly evident in long-lead finishing materials, imported systems, specialised fabrication items, and weather-sensitive construction activities where timing of availability becomes critical to maintaining continuity of execution.

Construction disruption frequently originates not from catastrophic events, but from relatively small interruptions occurring at operationally sensitive moments within the project lifecycle.

3.3 Approval Delays and Secondary Impact

Approval delays are frequently treated as purely time-related events.

In practice, however, their impact is rarely limited to delay alone.

Delayed approvals often result in:

- fragmented work fronts;
- repeated resequencing;
- underutilised labour;

- interrupted continuity;
- and repeated mobilisation and demobilisation.

These effects progressively reduce execution efficiency.

The resulting productivity loss may continue even after the approval itself is finally issued. Consequently, secondary operational impacts are often more significant than the approval delay itself.

3.4 Inadequate or Disclaimed Site Data

Many projects involve site information accompanied by disclaimers stating that the data is "indicative only."

In practice, however, such information is often:

- incomplete;
- limited in coverage;
- or non-representative of actual conditions.

This is particularly significant in geotechnical environments where subsurface variability may change substantially across short distances.

Contractors are frequently expected to extrapolate partial data across the entire project area.

However, in many situations:

- independent verification is impractical;
- access constraints exist;
- regulatory approvals are required;
- or tender-stage timelines are insufficient.

The practical consequence is that:

risk allocation based solely on contractual disclaimers may not reflect actual execution feasibility.

3.5 Partial Possession and Logistics Constraints

Partial possession or phased handing over of completed sections is common in infrastructure and urban projects. Such arrangements frequently alter the Contractor's planned logistics and execution sequencing. It often disrupts the Contractor's planned site logistics. The resulting impacts include:

- loss of storage and staging areas
- restricted access routes
- interference with material movement
- fragmentation of work fronts These constraints result in:
- inefficient sequencing of activities
- increased material handling effort
- reduced labour productivity

Such conditions constitute a clear case of execution disruption, even if the overall completion timeline remains unaffected.

3.7 Vague Contractual Clauses

Certain contractual provisions are inherently ambiguous and open to interpretation. Common examples include:

- "The Contractor shall maintain sufficient stock of materials"
- "Data provided is for reference only"

Such clauses are often used to transfer risk to the Contractor without clearly defining:

- the extent of obligation
- measurable benchmarks
- practical limitations

In reality, terms such as "sufficient" must be interpreted in the context of:

- site constraints (space, logistics)
- industry practices (typical stocking cycles)
- commercial feasibility (cash flow and material handling)

Overly strict or retrospective interpretation of such clauses frequently leads to disputes and unjustified denial of claims.

3.8 Financial and Payment-Driven Disruption

In addition to physical and environmental factors, disruption in construction projects can also arise from financial constraints, particularly those linked to delayed payments. In many Government and Public Sector projects, Extension of Time (EOT) is often determined only:

- at or near the expiry of the contractual completion date; or
- after the milestone has already been exceeded.

The approval process itself may take an additional 2–3 months, during which:

- interim payments are withheld or delayed;
- Contractor bills are not processed pending EOT determination.

Working Capital Impact

This delay in payment places significant strain on the Contractor's working capital, leading to:

- delayed payments to subcontractors and vendors;
- cash flow imbalance across project operations;

The consequences at site level are immediate and tangible:

- subcontract labour may not report to work;
- drivers and operators may skip duties;
- supply chains become irregular;

This results in:

- reduced workforce availability;
- intermittent work progress;
- disruption to planned sequencing and productivity.

Nature of Disruption

Unlike traditional disruption caused by physical events, this form of disruption:

- originates from financial stress;
- manifests as resource instability;
- and leads to loss of continuity in execution.

Such impacts are particularly common in long-duration projects, where sustained cash flow is critical for maintaining workforce and supply chain stability. Limitations in Claiming Despite the clear impact, such disruption is rarely claimed due to several practical challenges:

- contracts often require Contractors to maintain minimum working capital, limiting entitlement;
- causation between delayed payment and productivity loss is difficult to establish objectively;
- Contractors may be reluctant to disclose financial stress or internal cash flow constraints.

Resource Loading and Proof Challenges A further complication arises from the absence of resource-loaded baseline programmes in many projects. As a result:

- Contractors are unable to demonstrate:

- planned vs actual manpower deployment;
- productivity loss due to reduced workforce;
- Employers and Engineers may counter-argue that:
- the Contractor was under-resourced;
- labour shortages are attributable to contractor inefficiency rather than external factors.

This creates situations where even genuine disruption cannot be effectively substantiated because of inadequate planning data. Industry Context — Labour Constraints The issue is further compounded by broader industry trends:

- widespread infrastructure development leading to labour scarcity;
- projects operating at 60–70% effective capacity;
- Contractors frequently operating in catch-up mode, rather than as per baseline schedule.

In such conditions:

- even minor disruptions have amplified impact;
- distinguishing between:
- systemic labour shortage; and
- project-specific disruption becomes increasingly difficult.

Emerging Risk Trend

An emerging industry pattern is:

- under-deployment of manpower;
- coupled with over-deployment of machinery;

This imbalance is often used by Employers to:

- deny EOT;
- reject disruption claims; on the basis that the Contractor has not optimally utilised available resources.

Key Insight

Financial stress-induced disruption is real and operationally significant, but often remains unclaimed due to contractual constraints, evidentiary challenges, and resource planning deficiencies

Chapter 4

Fit for Purpose and Duty of Care — Limits of Application

4.1 Introduction

In modern construction contracts, broad obligations such as “fit for purpose” and “duty of care” increasingly occupy a central position in risk allocation.

Employers and Consultants frequently rely upon these provisions while evaluating claims arising from:

- unforeseen site conditions;
- productivity loss;
- execution difficulties;
- design-related complications;
- and operational inefficiencies.

In many projects, these obligations are interpreted expansively and applied as broad risk-transfer mechanisms.

As a result, Contractors are often expected to:

- absorb execution inefficiencies;
- overcome incomplete information;
- manage unforeseeable operational conditions;
- and deliver the completed works irrespective of practical difficulties encountered during execution.

However, project realities demonstrate that such obligations cannot be interpreted in isolation from:

- reasonableness;
- foreseeability;
- adequacy of information;
- practical feasibility;
- and the actual conditions under which the work must be executed.

This chapter therefore examines the practical boundaries of these obligations and explores how their overextension frequently contributes to claim disputes and imbalance in risk allocation.

4.2 Nature of Fit for Purpose Obligations

The concept of fit for purpose fundamentally concerns the performance outcome of the completed works.

In essence, the Contractor is expected to deliver a finished product capable of achieving the intended function specified under the contract.

Examples may include:

- a bridge carrying specified traffic loads;
- a drainage system achieving required discharge capacity;
- or a treatment plant meeting defined operational standards.

The obligation therefore focuses primarily upon:

- functionality;
- operational performance;
- reliability;
- and achievement of intended use.

From a contractual perspective, this obligation serves an important commercial purpose.

Employers rely upon Contractors not merely to execute activities mechanically, but to ensure that the completed asset performs adequately in operation.

However, execution realities introduce complexities extending beyond theoretical performance obligations.

Construction is not carried out within controlled laboratory conditions.

Execution occurs within environments affected by:

- incomplete information;
- changing site conditions;

- environmental variability;
- safety constraints;
- and evolving operational realities.

A fit-for-purpose obligation cannot reasonably imply that the Contractor assumes unlimited responsibility for every uncertainty affecting execution.

4.3 Duty of Care and Reasonable Skill

The duty of care similarly requires the Contractor to exercise:

- reasonable professional competence;
- skill;
- diligence;
- and proper execution practices.

This includes obligations such as:

- appropriate planning;
- competent supervision;
- safe execution;
- quality control;
- and reasonable engineering judgement.

Importantly, however, the duty of care is not equivalent to absolute foresight.

A Contractor may act competently and diligently while still encountering:

- unforeseen subsurface conditions;
- hidden obstructions;
- abnormal environmental impacts;
- supply-chain disruptions;
- or productivity losses arising from Employer-driven changes.

The boundary ultimately lies in the concept of reasonableness.

The Contractor is expected to exercise the standard of a reasonably competent and experienced contractor operating under similar circumstances.

This standard does not require:

- prediction of every future contingency;
- assumption of undefined risk;
- or absorption of all inefficiencies arising during execution.

4.4 Practical Disconnect Between Contractual Theory and Execution Reality

One of the most common issues in construction claims arises from the disconnect between:

- contractual risk allocation;
- and actual execution feasibility.

Tender-stage assumptions are often developed using:

- limited geotechnical data;
- indicative drawings;
- incomplete access conditions;
- preliminary utility information;
- or conceptual sequencing assumptions.

Contractors are nevertheless expected to price and plan execution based upon these limited inputs.

In many situations, independent verification is either:

- commercially impractical;
- operationally impossible;
- or restricted by tender timelines and regulatory constraints.

This becomes particularly significant in marine, coastal, urban, and underground infrastructure projects.

For example:

- geotechnical investigations may cover only a small percentage of the actual alignment;
- underground utility mapping may be incomplete;
- or access for independent verification may itself require regulatory approvals.

Despite these limitations, Contractors are often expected to assume full execution risk based upon broad disclaimer provisions.

The resulting imbalance frequently becomes visible only during execution.

4.5 Misapplication of Broad Contractual Obligations

In practice, "fit-for-purpose" and "duty-of-care" provisions are sometimes applied as blanket defenses against claims.

This commonly occurs in cases involving:

- unforeseen ground conditions;

- productivity loss;
- weather-related disruption;
- supply-chain constraints;
- or changes in execution environment.

The underlying reasoning often becomes:

"The Contractor accepted responsibility for execution and therefore bears the consequences."

However, such interpretation oversimplifies construction realities.

Execution methodology is not developed in isolation. It depends fundamentally upon:

- the accuracy of available information;
- the continuity of access;
- sequencing assumptions;
- environmental conditions;
- and operational feasibility.

Where these assumptions materially change, the execution environment itself changes.

The Contractor may still remain obligated to complete the works. However, the manner in which the work is executed may become substantially less efficient.

This distinction is extremely important in disruption claims.

The issue is not whether the Contractor must ultimately perform the work. The issue is whether the work is being performed under conditions materially different from those reasonably contemplated at tender stage.

4.6 Reasonableness and Foreseeability

A practical and balanced interpretation of contractual obligations requires consideration of:

- what could reasonably be foreseen;
- what could practically be verified;
- and what risks could realistically be priced at tender stage.

Construction projects inherently involve uncertainty.

However, uncertainty does not automatically justify unlimited transfer of risk.

For example:

- a Contractor may reasonably anticipate seasonal rainfall;
- but may not reasonably foresee prolonged abnormal working restrictions caused by

cumulative weather patterns interacting with fragmented work fronts.

Similarly:

- a Contractor may reasonably accept some subsurface variability;
- but cannot reasonably extrapolate full-site conditions from severely limited geotechnical coverage where independent investigation is impractical.

The governing principle is that reasonable foreseeability defines the practical boundary of contractual risk.

4.7 Safety, Workability, and Execution Feasibility

An important practical issue frequently overlooked in claim assessment is the distinction between:

- structural capability;
- and execution feasibility.

Construction activities are not governed solely by design thresholds.

They are also governed by:

- human safety;
- workability conditions;
- labour movement;
- equipment operability;
- and continuity of execution.

For example, in coastal infrastructure projects:

- structural design wind speeds may remain well within permissible limits;
- yet actual site execution may become unsafe due to sustained wind exposure affecting labour stability and workability.

Similarly:

- working at height under moderate but continuous wind conditions may become operationally impossible;
- even though structural design thresholds are not exceeded.

The practical implication is significant.

Execution feasibility cannot be evaluated purely through design criteria.

The actual question is:

Can the work be carried out safely, continuously, and productively under the prevailing conditions?

This distinction frequently becomes central in disruption-related claims.

4.8 Relationship with Disruption Claims

The boundaries of fit for purpose obligations become particularly important when evaluating disruption.

Even where:

- the Contractor remains obligated to complete the works;
- no formal variation exists;
- or no Extension of Time is granted;

productivity loss may nevertheless arise because:

- execution continuity has been disturbed;
- working conditions have materially changed;
- or operational efficiency has deteriorated.

Fit-for-purpose obligations do not eliminate the possibility of disruption entitlement.

The Contractor may still validly argue that:

- the work had to be executed under materially different conditions;
- productivity reduced substantially;
- and additional effort became necessary to maintain progress.

The distinction between:

- obligation to complete;
- and obligation to absorb inefficiency;

therefore becomes critical.

4.9 Practical Insight — Limits of Obligation

This issue can be understood through a simple practical analogy.

A driver may be responsible for safely reaching a destination. However, that responsibility does not mean the driver must anticipate every unexpected road collapse, sudden traffic diversion, or obstruction that was neither visible nor reasonably foreseeable at the beginning of the journey.

The responsibility remains to complete the journey competently and safely.

But the existence of that responsibility does not eliminate the reality that:

- unexpected conditions may alter the route;
- increase fuel consumption;
- delay movement;
- or require additional effort.

Construction projects often operate similarly.

The Contractor remains responsible for delivering the completed works.

However, this responsibility cannot reasonably imply absorption of every unforeseen inefficiency or execution difficulty irrespective of:

- practicality;
- foreseeability;
- or adequacy of available information.

4.10 Conclusion

Fit for purpose and duty of care remain essential principles within construction contracts.

However, their application must remain aligned with:

- practical execution realities;
- reasonableness;
- foreseeability;
- and balanced risk allocation.

Overextension of these obligations frequently leads to:

- unjustified rejection of legitimate claims;
- distorted allocation of execution risk;
- and disputes arising from operational realities rather than contractual intent.

The practical lesson is therefore clear:

Fit for purpose ensures delivery of the intended outcome. It does not require unlimited absorption of unforeseeable execution inefficiencies..

Chapter 5

Commercial Realities and Behavioural Patterns in Construction Claims

5.1 Introduction

Construction claims are frequently analysed through legal and technical frameworks. However, in practice, Claim outcomes are often shaped equally — if not more strongly — by:

- commercial behaviour;
- organisational culture;
- negotiation dynamics;
- financial pressures;
- timing of decisions;
- and human response to risk.

Many claims that appear strong in principle ultimately fail due to:

- weak documentation;
- delayed action;
- internal organisational disconnect;
- reluctance to disclose information;
- or commercially defensive behaviour.

Similarly, some claims with uncertain entitlement are resolved commercially because:

- litigation risk exists for both parties;
- project continuation becomes more important than strict interpretation;

- or practical settlement becomes preferable to prolonged dispute.

Construction claims therefore cannot be understood purely as contractual exercises.

They are also reflections of:

- behavioural patterns;
- commercial pragmatism;
- organisational capability;
- and timing of decision-making.

5.2 Practical Insight — Early Warning Signals

This can be understood through a simple practical observation.

When milk is heated, it begins to rise gradually at the edges before boiling over. This acts as an early warning signal. If the heat is reduced at that stage, overflow can be prevented. If ignored, the milk eventually boils over, resulting in loss and disorder.

Construction projects often behave similarly.

Major disputes and claim situations rarely emerge suddenly.

They are usually preceded by smaller operational signals such as:

- minor productivity decline;
- intermittent labour shortage;
- delayed approvals;
- fragmented work fronts;
- emerging supply-chain instability;
- or increasing resequencing of activities.

Initially, these signs may appear manageable.

However, when they remain unaddressed over extended periods, they progressively develop into:

- larger disruption;
- critical-path delay;
- financial stress;
- and formal disputes.

The importance of early intervention therefore becomes significant.

Many project disputes escalated not because the original issue was unmanageable, but because the warning signals were ignored until the operational consequences became substantially more difficult to control.

The lesson is therefore straightforward:

Early intervention is often more effective than post-failure recovery.

5.3 Profit Erosion vs Actual Loss

One of the most important commercial realities in construction claims is the distinction between:

- actual financial loss;
- and erosion of anticipated profitability.

In many disruption situations, the Contractor may still complete the works without recording an accounting loss.

However, the Contractor may consume:

- substantially greater labour effort;
- higher equipment usage;
- increased supervision;
- or additional indirect cost.

The project therefore remains commercially viable, but the originally anticipated margin reduces significantly.

This creates an important practical challenge.

From an operational perspective, disruption is real. From an accounting perspective, however, the loss may not appear clearly identifiable.

For example:

- planned productivity = 100 units/day;
- achieved productivity = 70 units/day.

The Contractor may still complete the work through:

- additional labour;
- overtime;
- acceleration;
- or compressed sequencing.

The inefficiency exists operationally, but translating it into auditable financial loss becomes difficult.

This explains why many genuine disruption impacts remain:

- commercially under-recognised;
- inadequately substantiated;
- or partially unrecovered.

5.4 Reluctance Towards Open-Book Disclosure

Disruption claims often require detailed substantiation involving:

- productivity records;
- labour deployment data;
- equipment utilisation;
- and actual cost information.

This creates a commercial dilemma for Contractors.

To establish entitlement convincingly, substantial internal financial information may need to be disclosed.

However, Contractors are frequently reluctant to provide:

- internal cost structures;
- pricing methodology;
- productivity assumptions;
- or profit margins.

This reluctance arises because such disclosure may:

- weaken future negotiation positions;
- expose inefficiencies;
- or provide commercial leverage to the Employer.

As a result, Contractors sometimes:

- avoid pursuing disruption claims entirely;
- restrict claim scope;
- or attempt to frame disruption issues as simpler delay-related claims.

As a result, commercial sensitivity often limits the effectiveness of otherwise valid claims.

5.5 Cost of Enforcement and Dispute Aversion

The cost of pursuing claims significantly influences Contractor behaviour.

Formal dispute resolution mechanisms such as:

- arbitration;

- adjudication;
- or litigation;

often involve:

- substantial legal cost;
- expert fees;
- prolonged timelines;
- management effort;
- and uncertainty of outcome.

Contractors therefore frequently assess claims not merely on entitlement, but on:

- commercial recoverability;
- enforcement cost;
- and probability of success.

This creates a behavioural hierarchy in claim pursuit.

Lower-value claims are often:

- absorbed;
- commercially compromised;
- or abandoned.

Only claims of sufficient commercial significance typically proceed into formal dispute mechanisms.

An important reality is that entitlement alone does not determine whether a claim will actually be pursued.

5.6 Negotiation Versus Strict Entitlement

In practical project environments, many claims are ultimately resolved through commercial negotiation rather than strict legal determination.

This occurs particularly where:

- entitlement exists but quantification is uncertain;
- causation is arguable rather than definitive;
- or both parties face litigation risk.

Partial settlements therefore become common.

Such settlements frequently reflect:

- practical compromise;

- commercial continuity;
- avoidance of prolonged disputes;
- and shared uncertainty.

The resulting settlement may not fully align with strict contractual entitlement.

However, it may still represent the most commercially rational outcome for both parties.

Construction claims therefore operate within both:

- legal frameworks;
- and commercial ecosystems.

5.7 Internal Organisational Constraints

An often-overlooked aspect of construction claims is the Contractor's internal limitation.

Many Contractors face difficulties due to:

- weak documentation culture;
- absence of structured productivity tracking;
- fragmented reporting systems;
- or disconnect between project teams and head office.

In some cases, management may even avoid producing data that weakens the company's commercial position.

This creates situations where:

- operational impacts are known internally;
- but objective substantiation remains unavailable.

The practical consequence is significant.

Even valid claims may weaken because:

- data is incomplete;
- records are inconsistent;
- or contemporaneous evidence is absent.

This highlights an important principle:

claim strength depends not only upon entitlement, but also upon organisational capability.

5.8 Practical Insight — Decision Timing

This can be understood through a simple practical observation.

At a traffic signal, movement is governed by clearly defined phases—stop, prepare, and proceed. Ignoring these signals creates confusion, inefficiency, and conflict.

Construction claims operate in a similar manner.

Different stages of project execution require different forms of timely decision-making.

For example:

- notice stage → issue timely notification;
- impact stage → assess operational consequences;
- evaluation stage → determine entitlement and strategy.

Problems frequently arise not because the issue itself lacks merit, but because decisions are delayed until the situation becomes substantially more difficult to manage.

Late identification of:

- delay impact;
- disruption patterns;
- or productivity loss;

often weakens otherwise valid claims.

Similarly, delayed evaluation by Employers and Consultants frequently results in:

- accumulation of unresolved issues;
- fragmented records;
- and retrospective reconstruction of events.

The key lesson is that timely decisions are often as critical as correct decisions.

5.9 EOT Determination Practices and Deferred Evaluation

In many projects, particularly within public-sector environments, Extension of Time assessments are not carried out contemporaneously.

Instead, EOT evaluation is often deferred until:

- the contractual completion date approaches;
- milestone dates are exceeded;
- or project completion becomes commercially unavoidable.

During earlier stages, Consultants frequently adopt strict contractual positions while rejecting claims based upon:

- notice provisions;
- narrow clause interpretation;
- or incomplete preliminary analysis.

However, as the project progresses, practical realities become increasingly difficult to ignore.

Multiple delay events accumulate. Sequences overlap. Actual execution diverges substantially from original planning assumptions.

At this stage, retrospective analysis becomes increasingly complicated.

Consultants may then request:

- detailed delay analysis;
- as-built reconstruction;
- productivity data;
- and critical path assessment.

However, many Contractors are unable to produce fully compliant retrospective analyses because:

- contemporaneous programme updates were weak;
- resource loading was absent;
- or project records were inconsistently maintained.

The result is often a shift from:

- structured contractual evaluation;
- toward subjective commercial judgement.

This creates several consequences:

- inconsistency between initial rejection and final recommendation;
- increased audit exposure;
- reduced transparency;
- and greater dispute potential.

A more effective approach would involve:

- contemporaneous evaluation;
- running assessment of delay events;
- and continuous updating of EOT position throughout the project lifecycle.

Such an approach improves:

- transparency;
- audit defensibility;
- and overall fairness in claim administration.

5.10 Financial Stress and Behavioural Distortion

Financial stress exerts a powerful influence on project behaviour.

Delayed certification and payment frequently create:

- subcontractor instability;
- labour attrition;
- interrupted supply chains;
- and reduced workforce continuity.

These effects may not immediately appear within formal programme analysis.

However, operationally, they create substantial disruption.

Projects gradually begin operating in:

- fragmented sequences;
- catch-up mode;
- or survival-oriented execution environments.

Under such conditions:

- even minor operational disturbances create amplified impact;
- labour availability fluctuates unpredictably;
- and productivity becomes increasingly unstable.

Importantly, Contractors may avoid openly discussing such financial pressures because:

- disclosure weakens negotiation position;
- contractual provisions may require minimum financial capability;
- or reputational concerns exist.

As a result:

financially driven disruption often remains visible operationally but under-claimed formally.

5.11 Organisational Capability and Data Culture

Modern projects generate substantial volumes of operational data.

However, many organisations fail to systematically:

- structure;
- preserve;
- analyse;
- and utilise this information effectively.

Project teams often focus primarily upon:

- immediate execution;
- monthly billing;
- and short-term operational targets.

Meanwhile, head-office commercial teams may focus primarily upon:

- dispute positioning;
- financial exposure;
- or arbitration strategy.

This disconnect frequently causes:

- fragmented claim development;
- loss of contemporaneous evidence;
- and inconsistent commercial strategy.

A strong claims culture therefore requires integration between:

- project execution;
- documentation systems;
- planning functions;
- and commercial management.

Increasingly, digital systems provide opportunities for:

- real-time progress tracking;
- structured productivity monitoring;
- integrated document management;
- and traceable decision records.

However, technology alone is insufficient.

The real challenge remains behavioural:

whether organisations consistently capture and preserve information with long-term claim defensibility in mind.

5.12 Conclusion

Construction claims are shaped not only by contractual rights, but also by behavioural patterns, commercial pressures, and organisational capability.

Practical experience demonstrates that claim outcomes are strongly influenced by:

- timing of decisions;
- internal coordination;

- documentation culture;
- negotiation dynamics;
- and financial realities.

Many claims fail not because entitlement is absent, but because:

- warning signs were ignored;
- records were weak;
- strategy was inconsistent;
- or action was delayed.

The broader lesson is clear:

construction claims are as much behavioural and commercial processes as they are contractual exercises.

Successful claim management therefore requires alignment between:

- technical execution;
- commercial strategy;
- organisational systems;
- and timely decision-making throughout the project lifecycle.

Chapter 6

Case Illustrations — Practical Realities Behind Construction Claims

6.1 Introduction

Construction claims are often discussed in theoretical terms through reference to:

- contractual clauses;
- legal principles;
- delay methodologies;
- and formal entitlement frameworks.

However, actual project disputes rarely evolve in a purely theoretical manner.

In practice, claims arise from the interaction between:

- tender assumptions;
- execution realities;
- commercial pressures;
- incomplete information;
- behavioural responses;
- and operational constraints.

The following case illustrations reflect recurring patterns commonly observed across construction projects.

These cases demonstrate that:

- disruption often develops cumulatively rather than through isolated events;

- productivity loss frequently arises from indirect operational interference;
- contractual entitlement alone does not guarantee successful recovery;
- and claim outcomes are heavily influenced by timing, strategy, documentation, and commercial behaviour.

The objective of these illustrations is not merely to present disputes, but to examine:

- why the disputes emerged;
- how the claims were framed;
- where strategic mistakes occurred;
- and what broader practical lessons can be derived.

Collectively, these cases reveal a recurring theme:

Construction claims are shaped as much by execution realities and commercial conduct as by contractual language itself.

6.2 Case 1 — Basement Expansion and Soil Protection

In a development project involving a multi-level basement structure, the tender documentation provided only limited information regarding the extent and configuration of the basement works.

The tender drawings primarily indicated:

- overall plot boundary;
- built-up area;
- and general development parameters.

However, the actual extent of excavation and basement spread was not clearly identifiable from the tender-stage information.

Based on the available documents, the Contractor proceeded with pricing and planning using assumptions that appeared reasonable at tender stage.

During execution, it emerged that the basement footprint extended to approximately 80% of the plot area.

This materially altered the excavation geometry and changed the execution environment significantly.

The revised geometry created multiple technical and operational complications, including:

- substantially reduced setback from adjacent properties;
- increased risk of soil instability and collapse;

- restrictions on excavation sequencing;
- and increased safety concerns associated with deep excavation near adjoining structures.

As execution progressed, it became evident that conventional excavation support measures would not be adequate.

Specialised soil retention and stabilisation systems became necessary, including:

- soil nailing;
- temporary retention systems;
- and enhanced excavation protection measures.

The Contractor submitted a claim seeking recovery of the additional cost arising from these protective works.

The Employer initially rejected the claim on the basis that:

- excavation stability formed part of the Contractor's general execution obligation;
- and the Contractor remained responsible for ensuring safe construction.

However, the Contractor subsequently demonstrated that:

- the actual basement configuration materially differed from the tender-stage understanding;
- the extent of excavation significantly altered the practical risk profile;
- and the specialised retention measures were not reasonably foreseeable based upon the information provided during tender.

The Contractor further established that:

- the original pricing assumptions had been developed on the basis of substantially different excavation geometry;
- and the revised conditions fundamentally changed the execution methodology.

Following prolonged negotiation, the parties agreed to a partial commercial settlement.

Although the Contractor did not achieve full recovery, approximately 50% of the claimed cost was accepted.

Key Lessons

1. Execution methodology is heavily influenced by geometry and physical configuration.
2. Tender assumptions must be assessed in the context of the actual information reasonably available at bid stage.
3. Even where broad execution obligations exist, materially altered physical conditions

may justify additional entitlement.

4. Commercial settlement frequently emerges where strict contractual interpretation fails to fully address execution realities.

Key Insight

Changes in physical configuration can fundamentally alter execution methodology, even where the technical scope appears broadly similar.

6.3 Case 2 — Seasonal Material Availability and Approval Delay

In another project involving high-end architectural finishes, a critical granite material was required from a specific quarry source.

The material had been selected because:

- colour consistency was important;
- replacement sources were commercially unsuitable;
- and matching material availability from alternative suppliers was limited.

The Contractor was aware that the quarry historically became inaccessible during monsoon periods due to:

- flooding;
- waterlogging;
- and transportation restrictions.

Accordingly, the Contractor developed the procurement strategy around this seasonal limitation and planned to secure the material before the onset of the monsoon.

The material sample was submitted to the Employer for approval within the planned timeframe.

However, the approval process became substantially delayed.

Repeated follow-ups were undertaken, but the final approval was issued only after the commencement of the monsoon season.

By that stage:

- quarry operations had slowed significantly;
- access roads were affected by waterlogging;
- and extraction activities became practically restricted.

As a consequence:

- procurement timelines collapsed;

- downstream finishing activities became disrupted;
- and sequencing across multiple areas was affected.

The impact extended beyond mere delay.

The Contractor experienced:

- intermittent labour utilisation;
- fragmented execution of finishing activities;
- repeated remobilisation;
- and inefficient deployment of supervisory resources.

Although the Employer considered limited Extension of Time entitlement for the delayed approval itself, the wider productivity-related consequences were not initially recognised.

The resulting disruption was substantial.

The Contractor had originally planned:

- continuous execution;
- coordinated sequencing;
- and uninterrupted material flow.

Instead, the project entered a fragmented execution environment characterised by:

- intermittent availability of work fronts;
- labour inefficiency;
- and resequencing of interconnected activities.

The dispute highlighted an important distinction.

The primary issue was not merely delay in approval. The real operational consequence arose because the approval delay caused the Contractor to miss a critical procurement window governed by seasonal constraints.

Key Lessons

1. Employer delays interacting with seasonal constraints can create system-wide disruption.
2. Missed procurement windows frequently generate indirect productivity loss extending far beyond the original delay event.
3. Disruption often arises from interaction between events rather than isolated causes.
4. Procurement sequencing forms a critical component of practical project planning.

Key Insight

A delayed approval may appear administrative in nature, but when linked to seasonal supply constraints, it can fundamentally disrupt the entire execution sequence.

6.4 Case 3 — Scope Reduction and Inflated Subcontractor Claim

In a subcontract arrangement relating to specialised works, the subcontractor ultimately executed only approximately 27% of the originally anticipated scope.

The reduction arose due to:

- project design changes;
- revised execution strategy;
- and modification of overall project requirements.

The subcontractor subsequently submitted a substantial commercial claim asserting entitlement arising from the reduction in allocated work.

The claim value approached nearly 50% of the original subcontract value.

However, the claim narrative relied largely upon broad assertions rather than measurable substantiation.

The Contractor requested supporting details including:

- identification of fixed costs;
- unrecovered mobilisation expenditure;
- plant commitments;
- labour retention impact;
- and quantification of actual loss.

The subcontractor was unable to produce:

- structured cost breakdowns;
- audited supporting data;
- or objective analysis demonstrating actual commercial damage.
- Instead, the claim appeared to rely more upon perceived commercial expectation than demonstrable financial loss.

As discussions progressed, it became increasingly apparent that:

- while some entitlement arising from scope reduction may arguably have existed;
- the claimed amount lacked analytical foundation.

The absence of substantiation substantially weakened the credibility of the claim.

Ultimately, the subcontractor did not pursue the matter further.

Key Lessons

1. Reduction in scope does not automatically justify arbitrary compensation percentages.
2. Commercial expectation is not equivalent to demonstrable financial loss.
3. Claims require measurable substantiation rather than broad assertions.
4. Quantification remains as important as entitlement.

Key Insight

A claim may appear commercially understandable, but without analytical substantiation, practical recoverability becomes extremely limited.

6.5 Case 4 — Internal Data Constraints and Scope Dispute

In another project, the Employer proposed recovery against the Contractor on the basis that certain portions of the contractual scope had allegedly not been executed.

The Contractor disputed the Employer's interpretation and maintained that the scope had substantially been fulfilled.

However, the disagreement gradually exposed a deeper organisational issue.

A practical strategy was proposed to objectively analyse the dispute by comparing:

- tender-stage estimated quantities;
- against actual executed quantities.

Such analysis would likely have provided:

- measurable clarity regarding the executed scope;
- objective evidence supporting either position;
- and a stronger factual basis for commercial negotiation.

However, the Contractor's internal management did not provide the required data.

The reluctance to disclose or compile the information gradually weakened the Contractor's position.

The absence of supporting records created a strong commercial inference that:

- the actual executed quantities were probably lower than the tender assumptions.

As negotiations progressed, the Contractor's defence became increasingly dependent upon:

- contractual interpretation;
- general arguments;
- and negotiation posture;

rather than objective evidence.

The case illustrated that entitlement arguments alone are insufficient when reliable internal records are absent.

Key Lessons

1. Internal data availability frequently determines practical claim strength.
2. Absence of transparency weakens commercial credibility.
3. Organisational reluctance to disclose information can undermine otherwise arguable positions.
4. Objective quantity comparison often provides stronger support than purely contractual argument.

Key Insight

Claims are not weakened only by adverse facts—they are often weakened by absence of reliable internal information.

6.6 Case 5 — Value Engineering, Time Benefit, and Misaligned Incentives

In a major infrastructure project, the Contractor proposed an alternative execution methodology intended to improve:

- constructability;
- execution efficiency;
- and overall project duration.

The proposal involved a technically innovative design approach that had not previously been implemented within that particular project environment.

Initially, the Employer expressed hesitation because:

- the methodology was unfamiliar;
- execution risk appeared uncertain;
- and the design departed from conventional practice.

Following extensive technical deliberation, however, the proposal was accepted as a Value Engineering variation.

The acceptance was subject to a condition that:

- any realised savings would be shared between the parties.

At proposal stage, the Contractor indicated:

- modest direct cost savings;
- and approximately three months of time reduction.

However, the deeper commercial reality was considerably more complex.

The alternative methodology was not actually cheaper in direct execution terms.

In fact:

- execution cost under the revised approach was likely higher;
- additional engineering and operational effort became necessary;
- and implementation risk remained substantial.

The real advantage of the proposal lay primarily in:

- acceleration of completion;
- reduction of overall project duration;
- and avoidance of extended project exposure.

As execution progressed, the actual time benefit significantly exceeded original expectations.

The revised methodology ultimately reduced the overall duration by more than one year compared to the original execution strategy.

The resulting project-level benefits were substantial.

The Employer gained:

- earlier project utilisation;
- reduced escalation exposure;
- improved operational commencement;
- and significant time-related advantage.

The Contractor similarly benefited through:

- reduced prolongation exposure;
- lower preliminary overhead duration;
- and mitigation of future delay-related risk.

However, the subsequent commercial evaluation created dispute.

The Employer's Representative assessed the Value Engineering outcome primarily through:

- estimated cost comparisons;
- and theoretical savings calculations.

The Contractor, however, submitted actual cost records demonstrating that:

- the revised methodology incurred higher direct execution cost;
- and the original proposal had been commercially viable primarily because of time

benefit rather than cost reduction.

A central commercial argument subsequently emerged.

The Contractor contended that:

- once the proposal had been formally accepted as a contractual variation;
- valuation should be based upon actual execution cost;
- rather than theoretical estimated savings.

Following prolonged negotiation, the matter was ultimately resolved through commercial settlement.

The case exposed a recurring issue in Value Engineering assessments.

Traditional evaluation frameworks often recognise:

- direct cost reduction;

but inadequately recognise:

- time value;
- risk mitigation;
- acceleration benefit;
- and avoidance of future exposure.

An additional organisational issue also emerged.

Within the Contractor's own organisation, the initiative was not necessarily viewed positively because:

- it reduced the overall contract value;
- rather than increasing measurable revenue.

This reflected a broader behavioural pattern within construction organisations where: avoidance of loss is often not recognised as value creation.

Key Lessons

1. Time value frequently remains undervalued in Value Engineering evaluation.
2. Project benefit cannot always be assessed through direct cost comparison alone.
3. Actual execution cost may differ substantially from theoretical estimated savings.
4. Strategic risk reduction itself creates commercial value.
5. Organisational metrics often fail to recognise value preservation and mitigation.

Key Insight

Value Engineering should be evaluated through integrated assessment of:

- cost;

- time;
- risk;
- and project-wide benefit.

6.7 Case 6 — Wind Conditions, Safety Constraints, and Misapplication of Design Criteria

In a coastal infrastructure project, prolonged periods of high wind affected execution activities along exposed marine work fronts.

The wind conditions persisted continuously for approximately fifteen days.

Although the recorded wind speeds did not qualify as extreme weather under the contractual thresholds, the actual site conditions became operationally difficult.

The sustained wind exposure created:

- unsafe working conditions;
- instability for labour movement;
- restrictions on handling equipment and materials;
- and difficulty in maintaining safe execution along exposed areas.

The site safety department ultimately issued warnings restricting labour deployment in exposed seafront locations.

As a consequence:

- planned activities slowed substantially;
- productivity reduced;
- and certain operations became intermittently suspended.

The Contractor sought Extension of Time on the basis that:

- the prevailing conditions rendered execution practically unsafe and inefficient.

However, the claim was rejected.

The Engineer relied upon contractual interpretation referencing:

- IRC design wind-speed criteria;
- and the argument that the recorded wind speeds remained below the structural design thresholds.

This reasoning exposed a fundamental misunderstanding.

The cited design criteria related to:

- structural design capability;
- not safe working conditions for labour execution.

The actual operational issue was not whether the structure could withstand the wind. The issue was whether workers could safely and productively execute the activities under sustained exposure conditions.

The duration of the event became especially important.

The disruption did not arise from a short extreme cyclone event. Instead, it arose from:

- sustained moderate conditions;
- continuous operational restriction;
- and prolonged productivity reduction.

An additional practical dimension further weakened the Contractor's position.

At the time the disruption occurred:

- the project was primarily within foundation-stage execution.

The actual productivity impact during foundation activities therefore remained comparatively manageable.

Had the same conditions occurred during:

- superstructure activities;
- deck slab execution;
- or working-at-height operations;

the operational impact would likely have been substantially greater.

This highlighted an important principle.

The same environmental condition may create vastly different levels of disruption depending upon:

- stage of execution;
- work-front intensity;
- and nature of activities underway.

The Contractor also missed an opportunity to pursue:

- a structured disruption claim based upon productivity loss and safety-driven work restrictions.

Key Lessons

1. Structural design criteria are not equivalent to safe execution criteria.
2. Sustained moderate conditions may create greater disruption than short extreme events.
3. Environmental impact varies significantly according to stage of project execution.

4. Safety restrictions can provide strong evidentiary support for disruption claims.
5. Productivity loss may remain claimable even where strict EOT entitlement is disputed.

Key Insight

Execution feasibility is governed not merely by design thresholds, but by the practical ability to safely and continuously perform the work.

6.8 Case 7 — Undisclosed Working Conditions and Mischaracterised Claims

In a coastal flood-protection project, the Contractor was required to construct deep sluice gate chambers integrated within an existing drainage system.

The chambers extended approximately six to eight metres below ground level.

At tender stage, the works appeared technically consistent with conventional stormwater infrastructure.

However, during execution, it became evident that:

- the drainage system also carried continuous domestic wastewater discharge;
- and the working environment differed substantially from what had reasonably been anticipated.

The Contractor encountered:

- continuous foul odour;
- unhygienic conditions;
- persistent wastewater flow;
- and severe discomfort for labour deployment.

The execution environment became even more complicated because:

- work could only proceed during low-tide windows;
- while high tide submerged substantial portions of the work area.

Even during workable periods:

- sewage flow persisted;
- labour morale deteriorated;
- productivity reduced substantially;
- and continuity of execution became difficult.

Despite these conditions, the Contractor did not contemporaneously frame the issue as:

- adverse working conditions;
- disruption;

- or health-and-safety-related productivity loss.

Instead, the Contractor attempted to recover commercial impact indirectly through:

- inflated quantity claims;
- rather than clearly articulating the actual operational difficulties.

This strategic approach weakened the overall credibility of the claim.

The Employer and Consultant treated the sluice gate works primarily as:

- a variation in physical scope;

without fully evaluating:

- the altered execution environment;
- labour-related productivity constraints;
- or the human and operational realities associated with the work.

The case demonstrated that the technical scope may remain similar while workability conditions differ fundamentally.

Key Lessons

1. Workability conditions can materially affect productivity even where technical scope remains unchanged.
2. Human and environmental realities form part of execution feasibility.
3. Claims must identify the real operational cause rather than indirect symptoms.
4. Failure to contemporaneously articulate disruption weakens entitlement.

Key Insight

Construction claims are influenced not only by what must be built, but by the conditions under which people must execute the work.

6.9 Case 8 — Inadequate Geotechnical Data, Time Bar, and Loss of Entitlement

In a major foundation project, the Employer provided geotechnical investigation data covering only approximately thirty percent of the required borehole information at tender stage.

The Contractor developed the piling design and execution assumptions based upon the available information.

However, the project environment created substantial practical limitations regarding independent verification.

The geotechnical investigation area was located within:

- a coastal and intertidal environment;

- subject to tidal restrictions;
- regulatory approvals;
- and limited physical accessibility.

Furthermore:

- the final alignment had not been fully established;
- exact borehole positioning remained uncertain;
- and tender-stage timelines did not realistically permit comprehensive independent investigation.

Under these conditions:

- meaningful independent verification was practically constrained.

Nevertheless, the contractual framework treated the geotechnical information as:

- indicative only.

During execution, the actual subsurface conditions differed materially from the tender assumptions.

Pile lengths increased by approximately thirty percent beyond the original basis.

This created:

- increased drilling duration;
- additional material consumption;
- equipment inefficiency;
- and delay affecting the critical path.

The Contractor initially pursued recovery for additional piling cost.

However, the Consultant rejected the claim on the basis that:

- the data was indicative;
- the Contractor bore interpretation responsibility;
- and fit-for-purpose obligations applied.

A more serious issue subsequently emerged.

Despite the critical-path impact:

- the Contractor failed to issue timely EOT notices;
- and did not preserve procedural entitlement under the contractual time-bar provisions.

As a consequence:

- subsequent attempts to recover time entitlement were rejected.

The Contractor ultimately became exposed to:

- milestone-related penalties;
- liquidated damages;
- and substantial unrecovered disruption impact.

Importantly, the Contractor also failed to independently pursue:

- productivity-based disruption entitlement arising from increased drilling effort and altered execution conditions.

The combined financial consequence became substantial.

The case illustrated a recurring practical problem in construction claims.

Even where:

- technical merit exists;
- and operational impact is genuine;

procedural failure can extinguish recoverability.

Key Lessons

1. Partial geotechnical data cannot reliably represent full-site subsurface conditions.
2. Practical feasibility of independent verification must form part of risk-allocation assessment.
3. Procedural compliance is critical for preserving entitlement.
4. Delay and disruption claims should be independently evaluated.
5. Strong entitlement may still fail due to weak claim administration.

Key Insight

In construction claims, entitlement is created by events—but preserved through timely action.

6.10 Case 9 — Post-Award Optimisation and Deferred Commercial Alignment

In a bridge infrastructure project, the Contractor originally proposed off-site precast segment fabrication at a yard located approximately forty kilometres from the project alignment.

The tender submission also mentioned the possibility of developing an on-site casting yard upon reclaimed land, although this option was not initially pursued.

Following contract award, the project encountered approximately nine months of delay due to litigation-related restrictions affecting reclamation activities.

Once the restrictions were lifted, it became apparent that:

- portions of the reclaimed area had stabilised sufficiently to support casting-yard development.

The Contractor proposed relocating the casting yard to the project site.

The revised arrangement offered several practical advantages including:

- reduced transportation distance;
- improved logistical coordination;
- and acceleration opportunity.

The Employer accepted the proposal subject to:

- regulatory approvals;
- and compliance obligations.

However, a critical commercial mistake occurred at proposal stage.

The Contractor did not:

- formally characterise the proposal as Value Engineering;
- quantify cost implications;
- identify time-saving benefits;
- or establish any commercial-sharing framework.

Similarly, the Employer also did not immediately address:

- savings assessment;
- benefit allocation;
- or revised commercial implications.

Execution proceeded.

Approximately six months later, the Employer raised a recovery claim asserting that:

- the Contractor had realised substantial commercial savings through:
- elimination of transportation cost;
- reduced land-rental expenditure;
- and improved operational efficiency.

The Contractor attempted to resist the recovery.

However, the defence remained commercially weak because:

- the original proposal lacked structured commercial definition;
- no quantified acceleration benefit had been established;
- and the Contractor had not documented how the revised arrangement mitigated delay exposure.

The Engineer subsequently assessed theoretical savings and initiated recovery.

Facing:

- cash-flow pressure;
- prolonged negotiation;
- and weakened bargaining position;

the Contractor ultimately accepted a negotiated settlement.

Although the nominal settlement represented approximately twenty-five percent of the Employer's assessment, the delayed recovery process created additional:

- financing burden;
- working-capital strain;
- and indirect commercial loss.

The case illustrated an important behavioural pattern.

Technically sound optimisation decisions can still create substantial commercial exposure when:

- contractual alignment;
- financial strategy;
- and negotiation positioning;

are not addressed simultaneously.

Key Lessons

1. Technical optimisation must be commercially structured at inception.
2. Delay mitigation and time benefit should be quantified contemporaneously.
3. Absence of commercial definition creates retrospective dispute risk.
4. Cash-flow impact significantly affects negotiation strength.
5. Delayed engagement frequently increases indirect commercial exposure.

Key Insight

Failure to commercially structure technical decisions at the time of implementation often results in loss of strategic control over later outcomes.

6.11 Consolidated Observations from the Case Studies

Across these case illustrations, several recurring patterns emerge.

First, many disputes originate from the gap between tender assumptions and actual execution conditions.

Second, disruption frequently develops through cumulative operational interference rather than isolated dramatic events.

Third, claim outcomes are heavily influenced by:

- timing of notices;
- quality of records;
- commercial positioning;
- and organisational behaviour.

Finally, the cases demonstrate that construction claims are shaped not only by contractual language, but also by execution realities, operational constraints, and commercial decision-making throughout the project lifecycle.

6.12 Conclusion

The case illustrations presented in this chapter demonstrate that construction claims emerge from the interaction between:

- contractual frameworks;
- physical execution realities;
- environmental conditions;
- organisational capability;
- and commercial behaviour.

The practical lessons extend beyond individual disputes.

Collectively, these cases reveal that:

- productivity loss is frequently indirect and cumulative;
- operational realities often differ materially from tender assumptions;
- and procedural strategy frequently determines practical recoverability.

Perhaps most importantly, the chapter demonstrates that:

successful claims are rarely determined by entitlement alone.

They depend equally upon:

- timing;
- framing;
- evidence;
- commercial judgement;

- and organisational preparedness.

The broader practical insight emerging from these case studies is therefore clear:

Construction claims are not isolated contractual events. They are reflections of how projects respond to uncertainty, change, pressure, and execution reality over time.

Chapter 7

Practical Framework for Disruption Claims

7.1 Introduction

Among all categories of construction claims, disruption claims remain one of the most difficult to establish, evaluate, and resolve.

Unlike conventional delay claims, disruption does not always manifest through visible extension of project duration.

Instead, disruption frequently appears through:

- reduced productivity;
- fragmented work fronts;
- inefficient sequencing;
- repeated mobilisation;
- increased effort for the same output;
- and loss of execution continuity.

In many projects, disruption may be obvious to site personnel. However, converting that reality into:

- structured entitlement;
- measurable causation;
- and defensible quantification;

remains extremely challenging.

A major reason for this difficulty is that disruption rarely arises from a single isolated event. More commonly, it develops progressively through:

- cumulative operational interference;
- interaction between multiple events;
- changes in working conditions;
- or prolonged inefficiency over extended periods.

Furthermore, disruption often affects:

- productivity;
- labour efficiency;
- resource utilisation;
- and workflow continuity;

without necessarily causing immediate visible delay to overall completion.

As a result, disruption claims are frequently:

- misunderstood;
- inadequately framed;
- weakly substantiated;
- or incorrectly merged into delay claims.

This chapter therefore presents a framework for identifying, structuring, and substantiating disruption claims.

The objective is not merely theoretical explanation, but development of a workable methodology aligned with:

- project realities;
- operational behaviour;
- evidentiary requirements;
- and commercial practicality.

7.2 Nature of Disruption

Disruption fundamentally relates to:

loss of productivity arising from interference with planned execution.

The essential characteristic of disruption is that:

- the same resources;
- under altered conditions;
- produce lower output than originally anticipated.

This may occur due to:

- fragmented work fronts;
- repeated stoppages;
- restricted access;
- inefficient sequencing;
- weather-related working difficulties;
- or operational constraints arising during execution.

Importantly, disruption does not always extend the project completion date.

A Contractor may:

- accelerate;
- deploy additional labour;
- increase equipment usage;
- or compress activities;

in order to maintain programme completion.

However, the underlying productivity loss still exists.

A project may therefore experience:

- severe disruption;
- substantial inefficiency;
- and major additional cost;

without necessarily generating equivalent Extension of Time entitlement.

Disruption must therefore be analysed primarily through:

- productivity behaviour;
- rather than merely programme duration.

7.3 Why Disruption Claims Frequently Fail

Despite genuine operational impact, disruption claims often fail in practice.

This occurs for several recurring reasons.

Absence of contemporaneous records

Many Contractors fail to systematically maintain:

- productivity data;
- resource deployment records;
- disruption logs;
- or planned-versus-actual comparisons.

As a result, the impact becomes difficult to demonstrate objectively.

Weak Causation

Disruption claims often rely upon broad narrative assertions without clearly linking:

- the disruptive event;
- to the resulting productivity impact.

This creates vulnerability during claim evaluation.

Mixing Delay and Disruption

Contractors often combine:

- delay;
- prolongation;
- acceleration;
- and disruption;

into a single undifferentiated claim.

This weakens analytical clarity.

Over-reliance on theoretical analysis

Claims sometimes depend excessively upon:

- formula-driven calculations;
- theoretical assumptions;
- or generic productivity factors;

without sufficiently demonstrating actual project-specific operational impact.

Delayed Claim Development

In many cases, disruption is recognised only after:

- productivity has already deteriorated significantly;
- records have become fragmented;
- and causation has become difficult to reconstruct.

As a result, disruption may be operationally real yet commercially unrecoverable due to inadequate claim structure.

7.4 Step 1 — Identification of the Triggering Event

Every disruption claim must begin with identification of the event or condition that interfered with planned execution.

The triggering event may include:

- delayed approvals;
- restricted access;
- partial possession;
- design changes;
- environmental constraints;
- abnormal weather impact;
- supply-chain interruption;
- or fragmented work sequencing.

The event must be:

- identifiable;
- reasonably definable;
- and distinguishable from ordinary Contractor inefficiency.

At this stage, the objective is not yet quantification.

The objective is to establish:

what changed.

A common mistake in disruption claims is immediate focus upon:

- cost;
- valuation;
- or compensation;

without first clearly identifying the operational interference itself.

The foundation of disruption analysis always begins with:

- recognition of the altered execution environment.

7.5 Step 2 — Demonstration of Changed Working Conditions

Once the triggering event is identified, the next requirement is to demonstrate how:

planned execution conditions materially changed.

This stage is critical because disruption does not arise merely from occurrence of an event.

It arises because the event forces the Contractor to work differently from what was originally planned.

Typical examples include:

- continuous work becoming fragmented;
- unrestricted access becoming restricted;

- sequential execution becoming resequenced;
- dry working conditions becoming wet and unstable;
- or planned material flow becoming intermittent.

The comparison therefore focuses upon:

Planned Conditions

As anticipated during:

- tender stage;
- baseline programme;
- methodology planning;
- and productivity assumptions.

Actual Conditions

As experienced during execution following the disruptive event.

This comparison establishes:

- the altered operational environment;
- and the resulting loss of execution efficiency.

A particularly important aspect is that:

changed working conditions may arise even without formal variation in technical scope.

The work itself may remain broadly similar. However, the manner in which the work must be executed becomes materially different.

7.6 Step 3 — Demonstration of Productivity Loss

The central component of a disruption claim is demonstrating that productivity reduced because of the changed conditions.

This requires objective demonstration that:

- the same resources;
- under disrupted conditions;
- produced lower output.

Examples may include:

- lower excavation output per equipment cycle;
- increased labour-hours for the same quantity of reinforcement;
- reduced paving continuity;
- or slower installation rates caused by fragmented access.

The productivity comparison may be established through:

- measured mile analysis;
- unaffected-versus-affected comparison;
- production-rate analysis;
- labour-hour comparison;
- or operational cycle evaluation.

For example:

Condition	Productivity
Planned / unaffected	100 units/day
Disrupted	65 units/day

The essential principle is that productivity loss must be demonstrated through operational evidence rather than broad narrative assertion.

However, practical limitations frequently arise.

Many projects lack:

- resource-loaded programmes;
- structured productivity tracking;
- or detailed daily production records.

In such situations, disruption analysis often requires:

- reconstruction from available site records;
- progress data;
- equipment logs;
- labour deployment reports;
- and contemporaneous correspondence.

The strength of the claim therefore depends significantly upon:

- quality of underlying data;
- and consistency of operational records.

7.7 Step 4 — Establishment of Causation

Perhaps the most important and most difficult aspect of disruption analysis is: causation.

The Contractor must demonstrate not merely that:

- productivity reduced;

but that:

- the reduction occurred because of the disruptive event.

This requires a logical chain connecting:

Event → Changed Conditions → Productivity Loss → Commercial Impact

Weak causation remains one of the most common reasons for failure of disruption claims.

Employers and Consultants frequently argue that:

- reduced productivity arose from Contractor inefficiency;
- under-resourcing;
- poor planning;
- or labour-management issues.

The Contractor must therefore isolate and demonstrate:

- the actual operational consequence arising from the disruptive event itself.

This requires:

- contemporaneous records;
- sequencing analysis;
- productivity comparison;
- and logical explanation.

An important principle emerges here.

The greater the delay between:

- occurrence of disruption;
- and identification of its impact;

the more difficult causation becomes to establish.

This reinforces the importance of:

- contemporaneous monitoring;
- early identification;
- and structured disruption tracking.

7.8 Step 5 — Quantification of Financial Impact

Once productivity loss and causation are established, the final step involves converting the operational impact into measurable financial terms.

This may include:

- additional labour cost;

- increased equipment utilisation;
- extended supervision;
- repeated mobilisation;
- additional site overheads;
- or inefficiency-related expenditure.

The objective is to identify:

- incremental cost arising specifically from the disruption.

Care must be taken to avoid:

- double counting;
- overlap with prolongation claims;
- or inclusion of unrelated inefficiencies.

In practice, quantification often becomes commercially sensitive because:

- open-book disclosure may be required;
- actual cost structures become visible;
- and internal inefficiencies may also emerge.

As a result, Contractors sometimes prefer:

- productivity-based approaches;
- normative comparisons;
- or incremental-cost methodologies.

Nevertheless, financial quantification must logically flow from the demonstrated disruption impact.

7.9 Practical Insight — Optimal Point of Closure

This issue can be understood through a simple practical observation.

During clarification of butter, heating produces a characteristic sound indicating that the process remains active. When the sound ceases, it signals that clarification is complete. Continuing to heat beyond that point results in over-burning and deterioration of quality.

Construction claims often follow a similar behavioural pattern.

During claim preparation:

- data is collected;
- analysis is refined;
- causation is evaluated;

- and narratives are progressively developed.

At a certain stage:

- sufficient evidence becomes available;
- the operational impact is reasonably established;
- and quantification reaches commercially workable clarity.

This becomes the appropriate point for:

- submission;
- negotiation;
- or strategic engagement.

However, many claims continue into excessive refinement.

This frequently leads to:

- over-analysis;
- delayed submission;
- dilution of key arguments;
- and increased procedural risk.

In some situations, the pursuit of perfect analysis ultimately weakens timely recovery.

The practical lesson is therefore important.

Effective claims require judgement not only in developing analysis, but also in recognising the appropriate stage to conclude and proceed.

7.10 Integration of Delay and Disruption

Delay and disruption frequently coexist within construction projects.

However, they remain analytically distinct.

Delay primarily concerns:

- project completion;
- critical-path impact;
- and time entitlement.

Disruption primarily concerns:

- productivity loss;
- execution inefficiency;
- and additional effort.

A single event may therefore generate:

- Extension of Time entitlement;
- prolongation impact;
- and disruption-related productivity loss simultaneously.

For example:

- delayed access may extend completion duration;
- while also causing fragmented execution and labour inefficiency.

The analytical challenge is therefore not separation of events, but separation of:

- time impact;
- and productivity-related inefficiency.

Failure to distinguish these elements frequently creates:

- overlap;
- duplication;
- inconsistent analysis;
- and weakened credibility.

A well-structured claim therefore maintains:

- clear analytical boundaries;
- while still recognising the practical interaction between delay and disruption.

7.11 Practical Challenges in Modern Projects

Modern projects create increasing difficulty in disruption analysis because:

- work fronts are highly fragmented;
- multiple contractors interact simultaneously;
- productivity varies dynamically;
- and execution environments change continuously.

Additionally, industry-wide issues such as:

- labour shortages;
- supply-chain instability;
- accelerated delivery expectations;
- and financial stress;

further complicate objective evaluation.

In many projects:

- baseline productivity itself may already be unstable.

This creates a difficult practical question:

Was productivity loss caused by the disruptive event, or by broader systemic inefficiency?

Accordingly, modern disruption claims increasingly require:

- integrated operational analysis;
- structured data systems;
- and continuous monitoring rather than retrospective reconstruction.

7.12 Conclusion

Disruption claims are fundamentally concerned with:

loss of execution efficiency arising from changed working conditions.

Their successful evaluation requires far more than broad narrative assertions.

A credible disruption claim must systematically demonstrate:

- the triggering event;
- the changed execution conditions;
- the resulting productivity loss;
- the causal linkage;
- and the corresponding financial impact.

Project experience demonstrates that:

- operational impact alone is insufficient;
- entitlement alone is insufficient;
- and theoretical analysis alone is insufficient.

Successful disruption claims emerge only where:

- operational reality;
- structured evidence;
- commercial judgement;
- and timely strategic action;

remain properly aligned.

Chapter 8

Key Principles Emerging from Construction Claims Practice

8.1 Introduction

Construction claims often appear highly project-specific.

Each dispute may involve:

- different contracts;
- different stakeholders;
- different technical conditions;
- and different commercial environments.

However, when practical claim experiences are examined collectively, several recurring principles consistently emerge.

These principles do not merely represent contractual observations. They reflect:

- behavioural realities;
- execution dynamics;
- and recurring patterns observed across construction projects.

The purpose of this chapter is therefore to consolidate the practical lessons emerging throughout the earlier discussions and case illustrations.

These principles do not merely represent contractual observations.

They reflect:

- behavioural realities;
- execution dynamics;
- and recurring patterns repeatedly observed across construction projects.

8.2 Principle 1 — Reasonable Risk Is Different from Unlimited Risk

Construction contracts frequently attempt to allocate extensive responsibility to Contractors.

However, an important practical distinction must always be maintained between:

- reasonable foreseeable risk;
- and unlimited undefined uncertainty.

Contractors are expected to manage:

- ordinary weather variations;
- routine productivity fluctuations;
- foreseeable execution challenges;
- and commercially manageable uncertainties.

However, an important distinction must always be maintained between:

- reasonably foreseeable risk;
- and unlimited undefined uncertainty.

In such situations, risk allocation cannot reasonably become absolute.

The practical boundary must remain linked to:

- foreseeability;
- reasonableness;
- and feasibility of assessment at tender stage.

This principle repeatedly appears in disputes involving:

- geotechnical uncertainty;
- environmental constraints;
- fragmented access;
- and hidden execution conditions.

Key Insight

Contractors accept risk as part of construction. They do not automatically accept unlimited uncertainty incapable of realistic evaluation or pricing.

8.3 Principle 2 — Disruption Is Fundamentally About Interference with Planned Execution

Disruption does not arise merely because an event occurs.

It arises because:

planned execution continuity is disturbed.

The operational consequence may include:

- fragmented work fronts;
- interrupted sequencing;
- repeated mobilisation;
- inefficient access;
- or reduced productivity.

A project may therefore continue progressing while still suffering substantial disruption.

This principle is extremely important because many projects focus excessively upon:

- completion dates;
- while under-recognising productivity deterioration.

In practice:

- work may continue;
- but at significantly reduced efficiency.

The true commercial impact therefore often lies not in visible stoppage, but in:

- inefficient execution over prolonged periods.

Key Insight

Disruption is not merely an event. It is the inefficient manner in which work is ultimately forced to be executed.

8.4 Principle 3 — Duration and Continuity Often Matter More Than Intensity

Construction claims often focus upon dramatic or high-intensity event.

However, practical project experience repeatedly demonstrates that:

moderate but sustained conditions often create greater disruption than short extreme events.

This principle appears clearly in:

- prolonged drizzle;
- sustained wind exposure;
- recurring fragmented access;
- or continuous operational interference.

A short severe event may:

- temporarily stop work;
- but permit relatively rapid recovery.

In contrast:

- prolonged moderate interference often continuously reduces productivity over extended durations.

The cumulative impact may therefore become substantially greater.

8.5 Practical Insight — Misleading Severity

This issue can be understood through a simple practical observation.

A short-duration heavy downpour may appear severe but cause only limited operational disruption. In contrast, continuous drizzle throughout the day may significantly disturb activities because of sustained interference with working conditions.

Construction projects often behave similarly.

Not all high-intensity events produce major disruption.

In many situations:

- moderate but continuous interference causes greater productivity loss;
- particularly where multiple work fronts remain active simultaneously.

For example:

- prolonged wet conditions may continuously affect earthwork mobility;
- repeated moderate access restrictions may fragment multiple activities;
- or sustained wind exposure may steadily reduce labour efficiency.

Impact also varies significantly with project stage.

The same event may create vastly different levels of disruption depending upon:

- number of active work fronts;
- intensity of concurrent execution;
- and stage of project progress.

For instance:

- a moderate disruption during early-stage foundation work may create limited impact because work fronts remain relatively few;
- whereas the same disruption during peak execution involving multiple interconnected fronts may create system-wide productivity loss.

Similarly:

- disruption affecting a critical activity near project completion may disproportionately affect commissioning and delivery commitments.

Impact assessment requires evaluation not merely of event intensity, but also:

- duration;
- continuity;
- timing;
- and operational context.

Key Insight

In construction claims, sustained operational interference frequently causes greater disruption than isolated high-intensity events.

8.6 Principle 4 — Contractual Clauses Cannot Be Interpreted in Isolation from Practical Reality

Clauses such as:

- "sufficient stock shall be maintained";
- "data is indicative only";
- or "no compensation for delay";

are often interpreted rigidly during disputes.

However, practical execution realities frequently reveal that:

- physical constraints;
- logistics;
- working-capital limitations;
- environmental conditions;
- and operational feasibility;

significantly influence how such clauses operate in practice.

Overly literal interpretation frequently ignores:

- practical construction methodology;
- industry norms;
- and commercial feasibility.

This often creates imbalance between:

- theoretical contractual wording;
- and realistic execution capability.

Key Insight

Contractual clauses must be interpreted within the context of practical execution realities rather than through isolated textual interpretation alone.

8.7 Principle 5 — Fit for Purpose Does Not Mean Absolute Liability

The obligation to deliver a functioning completed asset does not mean the Contractor must:

- foresee every uncertainty;
- absorb every inefficiency;
- or compensate for all externally induced disruption.

Construction projects inherently involve:

- incomplete information;
- changing site conditions;
- and evolving operational realities.

The boundary therefore remains linked to:

- reasonable foreseeability;
- professional diligence;
- and balanced risk allocation.

Key Insight

Fit for purpose ensures the completed outcome. It does not require unlimited absorption of unforeseeable operational inefficiency.

8.8 Principle 6 — Entitlement and Recoverability Are Not the Same

One of the most important practical lessons emerging from construction claims is the distinction between:

- entitlement in principle;
- and practical recoverability.

Many claims may possess:

- legitimate operational basis;
- contractual support;
- and measurable impact.

Yet recovery may still fail because of:

- weak documentation;

- delayed notice;
- poor quantification;
- commercial sensitivity;
- or procedural non-compliance.

This explains why:

many valid claims remain commercially unrealised.

The reality of claims therefore extends beyond legal entitlement.

Recoverability depends equally upon:

- timing;
- evidence;
- negotiation strength;
- and organisational preparedness.

Key Insight

A strong entitlement position may still fail if procedural and evidentiary foundations remain weak.

8.9 Principle 7 — Data Quality Determines Claim Strength

Construction claims ultimately depend heavily upon:

- contemporaneous records;
- productivity data;
- programme updates;
- resource tracking;
- and structured documentation.

Modern projects generate large volumes of operational information.

However, many organisations fail to:

- systematically preserve;
- structure;
- and utilise;

this information effectively.

As projects become increasingly complex, reliable data increasingly determines:

- credibility;
- causation strength;

- and dispute defensibility.

Key Insight

The quality of evidence frequently determines claim strength more than the theoretical strength of contractual argument.

8.10 Principle 8 — Claims Are Behavioural and Commercial Processes

Construction claims are not governed solely by contracts.

They are also shaped by:

- negotiation behaviour;
- organisational priorities;
- financial pressure;
- relationship management;
- and strategic decision-making.

Commercial settlement frequently reflects:

- pragmatic compromise;
- risk balancing;
- and practical continuation of projects.

The behaviour of:

- Contractors;
- Consultants;
- Employers;
- and project teams;

therefore significantly influences claim outcomes.

Key Insight

Construction claims are behavioural and commercial processes operating within contractual frameworks.

8.11 Principle 9 — Timely Action Preserves Entitlement

Many otherwise valid claims fail because:

- notices were delayed;
- disruption was not identified contemporaneously;
- records deteriorated;

- or causation became difficult to reconstruct.

The longer the delay between:

- occurrence of impact;
- and formal claim action;

the weaker practical recoverability often becomes.

This principle repeatedly emerges across:

- EOT disputes;
- disruption claims;
- and variation-related recovery.

Key Insight

Entitlement may originate from events. Recoverability depends upon timely recognition and action.

8.12 Conclusion

The principles emerging from practical construction claims reveal a consistent reality.

Successful claim outcomes depend not merely upon:

- contractual wording;
- or legal entitlement.

They depend equally upon:

- execution realities;
- operational behaviour;
- commercial judgement;
- documentation quality;
- timing of decisions;
- and organisational capability.

Across all project environments, one overarching lesson consistently emerges:

Construction claims are fundamentally reflections of how projects respond to uncertainty, disruption, operational pressure, and changing conditions throughout execution.

Understanding these principles therefore provides not merely legal awareness, but deeper practical insight into:

- how disputes arise;
- why claims succeed or fail;

and how balanced project administration can reduce conflict while improving commercial fairness.

Chapter 9

Future of Construction Claims — Towards a Data-Driven and Integrated Approach

9.1 Introduction

The construction industry is undergoing a significant structural transformation.

Projects are becoming:

- larger in scale;
- more technologically integrated;
- commercially more complex;
- and increasingly exposed to overlapping operational, environmental, financial, and regulatory risks.

At the same time, expectations regarding:

- transparency;
- accountability;
- productivity;
- governance;
- and dispute management;

continue to increase across both public and private sector projects.

Traditional approaches to construction claims were largely built around:

- retrospective analysis;
- fragmented documentation;
- manual records;
- and heavily narrative-driven evaluation.

While such methods evolved within earlier project environments, they are becoming less effective in modern construction ecosystems where:

- project interfaces are highly interconnected;
- execution cycles are compressed;
- and operational decisions occur in real time.

In many projects, claims are still developed only after:

- productivity has deteriorated;
- delay has accumulated;
- or disputes have escalated.

This reactive approach frequently results in:

- weak causation;
- incomplete records;
- inconsistent narratives;
- and avoidable conflict.

The future of construction claims therefore lies not merely in stronger contractual drafting, but in:

- integration of real-time data;
- structured documentation systems;
- digital project environments;
- and collaborative decision-making frameworks.

This chapter examines how construction claims are likely to evolve in response to:

- digital transformation;
- behavioural change;
- integrated project systems;
- and growing emphasis on evidence-driven project administration.

9.2 Shift from Reactive to Proactive Claim Management

Historically, construction claims have largely been treated as:

post-event recovery mechanisms.

Claims were commonly:

- identified late;
- reconstructed retrospectively;

- and developed after disputes had already emerged.

This created several recurring problems.

By the time claim preparation commenced:

- records were incomplete;
- personnel had changed;
- causation became difficult to establish;
- and operational realities had faded into fragmented recollection.

Modern project environments require a shift towards:

continuous and proactive claim management.

This means that:

- disruption events are identified contemporaneously;
- operational impacts are tracked in real time;
- and project teams continuously monitor evolving execution conditions.

Rather than viewing claims as isolated legal exercises, future project management systems will increasingly treat claims as:

integrated project-management functions.

Claim management will progressively become:

- predictive rather than reactive;
- operational rather than purely contractual;
- and evidence-driven rather than narrative-dependent.

Projects that successfully integrate:

- planning;
- progress monitoring;
- productivity tracking;
- and commercial administration;

will be better positioned to:

- identify emerging disruption;
- mitigate deterioration;
- and resolve disputes earlier.

9.3 Data as the Foundation of Future Claims

One of the most significant developments affecting construction claims is the recognition that:

project data itself is becoming a strategic asset:

project data itself is becoming a strategic asset.

Historically, construction records were often maintained primarily for:

- contractual compliance;
- progress reporting;
- or billing support.

However, modern project environments generate vast quantities of operational information capable of supporting:

- productivity analysis;
- disruption identification;
- resource optimisation;
- and dispute resolution.

These data sources include:

- digital progress records;
- equipment telemetry;
- labour deployment logs;
- drone mapping;
- environmental monitoring;
- schedule updates;
- and integrated communication systems.

When properly structured and analysed, such information can establish:

- chronology of events;
- operational constraints;
- productivity deterioration;
- and causal relationships.

Importantly, future claims will depend less upon:

- broad narrative explanation;

and more upon:

- structured operational evidence.

This transition fundamentally alters the nature of claim substantiation.

The strength of future claims will depend heavily upon:

- quality of data capture;
- consistency of project systems;
- and organisational capability to analyse operational information..

9.4 Building Information Modelling (BIM) and Integrated Digital Systems

Building Information Modelling (BIM) and related digital platforms are progressively transforming the way construction projects are:

- planned;
- coordinated;
- executed;
- and monitored.

Traditionally, project information existed in fragmented formats across:

- drawings;
- schedules;
- spreadsheets;
- correspondence;
- and isolated reporting systems.

This fragmentation frequently contributed to:

- inconsistent information;
- delayed decision-making;
- and difficulty establishing causation during disputes.

Integrated digital environments now allow increasing linkage between:

- design data;
- sequencing;
- progress;
- quantities;
- logistics;
- and operational constraints.

Future claims may therefore be supported through:

- time-linked construction simulations;
- digital sequencing analysis;
- integrated productivity tracking;

- and automated change identification.

This development significantly improves:

- transparency;
- traceability;
- and objectivity.

For example:

- access restrictions;
- sequencing changes;
- or interface conflicts;

may increasingly be demonstrable through integrated digital project records rather than retrospective narrative reconstruction.

Disputes that previously depended heavily upon:

- interpretation;
- memory;
- and subjective explanation;

may progressively become:

more evidence-based and analytically verifiable.

9.5 Artificial Intelligence and Predictive Analysis

Artificial Intelligence (AI) and advanced analytics are expected to influence:

- project management;
- productivity forecasting;
- risk assessment;
- and claim evaluation.

AI systems may progressively assist in:

- identifying emerging disruption patterns;
- analysing productivity trends;
- detecting schedule deviation;
- and correlating operational events with performance deterioration.

For example, future systems may automatically identify:

- repeated work fragmentation;
- abnormal productivity decline;

- resource instability;
- or supply-chain interruption;

before these issues evolve into major claims.

This creates the possibility of:

- earlier intervention;
- improved mitigation;
- and reduction of cumulative disruption.

However, an important practical caution also exists.

Construction projects involve:

- behavioural dynamics;
- commercial judgement;
- negotiation strategy;
- and evolving operational realities;

which cannot be fully reduced to algorithmic analysis.

Accordingly, while AI may significantly improve:

- analytical capability;
- pattern recognition;
- and operational visibility;

human judgement will remain central to:

- commercial decision-making;
- risk interpretation;
- and dispute resolution.

9.6 Digital Evidence and Future Dispute Resolution

As projects become increasingly digitised, dispute resolution itself is likely to evolve.

Historically, arbitration and litigation frequently involved:

- reconstruction of events;
- conflicting interpretations;
- and fragmented documentary evidence.

Future disputes may rely more heavily upon:

- structured digital records;
- integrated project timelines;

- real-time communication trails;
- and system-generated operational evidence.

This may significantly improve:

- objectivity;
- speed of evaluation;
- and evidentiary consistency.

Digital systems may also support:

- faster adjudication;
- improved technical understanding;
- and reduced reliance upon subjective recollection.

This evolution may also create broader institutional benefits.

Improved project data may contribute towards:

- stronger industry benchmarking;
- improved policy development;
- enhanced judicial understanding of construction disputes;
- and better future contract drafting.

9.7 Behavioural and Organisational Transformation

Technology alone cannot improve construction claims.

Many project failures arise not from absence of information, but from:

- organisational behaviour;
- fragmented communication;
- delayed decision-making;
- and weak documentation culture.

Accordingly, future claim management also requires behavioural transformation.

This includes:

- greater integration between project teams and commercial departments;
- improved awareness regarding claim preservation;
- stronger contemporaneous documentation culture;
- and earlier recognition of operational risk.

A recurring issue across projects is that:

site teams focus primarily upon immediate execution pressures, while commercial implications are addressed only after problems escalate.

Future project systems must therefore encourage:

- integrated operational and commercial awareness throughout execution.

9.8 Collaborative Contracting and Reduced Adversarialism

Traditional claim environments frequently become adversarial because:

- risk allocation is heavily one-sided;
- communication becomes defensive;
- and claims are treated as confrontational exercises.

However, increasingly complex projects require:

- collaboration;
- transparency;
- and shared operational problem-solving.

Future project environments may therefore increasingly adopt:

- collaborative contracting models;
- integrated project delivery systems;
- early warning mechanisms;
- and shared risk-management approaches.

Such systems encourage:

- earlier issue identification;
- joint mitigation;
- and reduction of cumulative disputes.

This does not eliminate claims.

However, it may significantly reduce:

- escalation;
- adversarial positioning;
- and destructive dispute cycles.

9.9 Practical Insight — Early Warning Signals

This issue can be understood through a simple practical observation.

When milk is heated, it begins rising gradually along the edges before boiling over. This serves as a clear early warning. If heat is reduced at that stage, overflow can be prevented.

If ignored, the milk inevitably boils over, creating loss and disorder.

Construction projects often exhibit similar patterns.

Disruption rarely emerges suddenly without warning.

Early signals often appear through:

- minor productivity decline;
- initial approval delays;
- labour instability;
- fragmented sequencing;
- or recurring operational inefficiency.

If these warning signs are recognised early:

- corrective measures;
- mitigation;
- resequencing;
- and commercial intervention;

may significantly reduce escalation.

However, when ignored:

- productivity deterioration accumulates;
- critical-path impact emerges;
- financial pressure intensifies;
- and disputes become increasingly difficult to resolve.

Effective project administration depends not merely upon solving major problems after escalation, but upon recognising and responding to early indicators before disruption becomes systemic.

Key Insight

Early intervention is frequently more effective and commercially valuable than post-failure recovery.

9.10 Future Role of Construction Professionals

The future construction professional will increasingly require:

- multidisciplinary capability;
- digital literacy;
- operational understanding;

- and commercial awareness.

Traditional separation between:

- engineering;
- planning;
- quantity surveying;
- commercial administration;
- and claims management;

is becoming less sustainable.

Future professionals must understand:

- how operational events affect productivity;
- how data supports claims;
- how behavioural decisions influence disputes;
- and how commercial strategy interacts with project delivery.

This evolution will likely create:

- more integrated project teams;
- stronger evidentiary systems;
- and improved claim defensibility.

9.11 Conclusion

The future of construction claims lies not merely in:

- stronger contractual drafting;
- or more complex legal mechanisms.

It lies in:

- integration of operational data;
- proactive project management;
- collaborative decision-making;
- and structured digital systems.

As projects become more technologically advanced and operationally interconnected, successful claim management will depend upon:

- real-time visibility;
- structured evidence;
- behavioural awareness;

- and integrated commercial strategy.

The projects most capable of managing future claims effectively will not necessarily be those with the most aggressive contractual positions.

Rather, they will be those capable of:

- recognising disruption early;
- maintaining reliable operational data;
- aligning technical and commercial decision-making;

and responding proactively before operational inefficiency escalates into major dispute.

Chapter 10

Final Reflections on Construction Claims and Project Realities

10.1 Introduction

Construction claims are often viewed narrowly as:

- contractual disputes;
- legal arguments;
- or post-project recovery mechanisms.

However, project experience consistently demonstrates that construction claims represent something much broader.

They reflect:

- how projects respond to uncertainty;
- how organisations behave under operational pressure;
- how decisions are made;
- how risks are allocated;
- and how commercial realities interact with engineering execution.

Throughout this book, multiple recurring themes have emerged.

These themes reveal that construction claims are rarely isolated contractual events.

Instead, they are usually the cumulative outcome of:

- operational interference;
- fragmented decision-making;
- incomplete information;
- delayed action;
- behavioural misalignment;

- and evolving execution realities.

The purpose of this final chapter is not to introduce new concepts, but to consolidate the broader practical understanding emerging from the earlier discussions and case studies.

10.2 Construction Projects Operate in Dynamic Environments

One of the most important realities of construction projects is that:

execution environments continuously evolve.

Projects rarely proceed exactly as originally planned.

During execution:

- conditions change;
- priorities shift;
- external influences emerge;
- approvals are delayed;
- logistics evolve;
- and operational constraints interact unpredictably.

Accordingly, claims frequently arise not because:

- one party deliberately acts improperly;

but because:

- original assumptions progressively diverge from actual execution reality.

This divergence may involve:

- physical conditions;
- productivity assumptions;
- sequencing;
- labour availability;
- environmental constraints;
- or financial pressure.

Construction claims must therefore be understood not merely as legal disagreements, but as reflections of:

changing operational environments.

10.3 The Gap Between Contractual Theory and Execution Reality

Modern construction contracts increasingly attempt to allocate risk comprehensively.

Through:

- disclaimers;
- broad obligations;
- strict notice requirements;
- and extensive risk-transfer clauses;

contracts often seek to define responsibility with increasing precision.

However, practical execution frequently reveals that:

construction environments cannot be fully reduced to contractual language alone.

Many operational realities remain:

- uncertain;
- evolving;
- or incapable of complete anticipation at tender stage.

This creates a recurring gap between:

- theoretical contractual allocation;
- and practical execution feasibility.

The case studies examined throughout this book repeatedly illustrate this principle.

Examples include:

- incomplete geotechnical information;
- undisclosed execution conditions;
- weather-related productivity loss;
- safety-driven operational constraints;
- and fragmented work environments.

These situations demonstrate that project realities often extend beyond simplistic contractual interpretation.

10.4 Productivity Loss Remains One of the Least Understood Project Risks

A recurring theme across construction disputes is the under-recognition of:

productivity deterioration.

Projects often focus heavily upon:

- delay;
- milestones;
- and completion dates.

However, substantial commercial impact often arises not through visible stoppage, but through:

continuous inefficiency.

This may include:

- repeated mobilisation;
- fragmented access;
- operational interference;
- labour instability;
- and cumulative disruption.

Importantly:

projects may continue progressing while productivity steadily deteriorates.

This creates a dangerous practical misconception.

Because work continues visibly, the underlying disruption may remain under-recognised until:

- cost escalation accumulates;
- profitability erodes;
- or recovery becomes impossible.

The practical lesson is significant.

Construction claims must therefore recognise:

- productivity behaviour;
- operational continuity;
- and cumulative inefficiency;

as central components of project performance.

10.5 Time Value Is Frequently Underestimated

Another recurring theme emerging across practical claim environments is the underestimation of:

time value.

Traditional claim assessment often focuses primarily upon:

- direct cost;
- quantities;
- or measurable savings.

However, construction projects derive enormous value from:

- earlier completion;
- reduced exposure to escalation;
- faster operational utilisation;
- reduced financing burden;
- and avoidance of future risk.

This becomes particularly visible in:

- Value Engineering;
- acceleration;
- and mitigation-related decisions.

In many projects:

- cost savings are recognised;
- while time savings remain undervalued.

This frequently results in:

- distorted commercial evaluation;
- misunderstanding of mitigation measures;
- and disputes regarding valuation.

Project value cannot be assessed solely through direct cost comparison.

True project value often includes:

- time;
- continuity;
- risk reduction;
- and operational flexibility.

10.6 Behavioural Factors Influence Claims as Much as Contracts

Behavioural patterns significantly influence:

- claim development;
- negotiation;
- and recovery.

These behavioural influences include:

- reluctance to disclose data;
- defensive project administration;

- delayed decision-making;
- fragmented communication;
- and organisational misalignment.

Claims therefore evolve not merely through contractual entitlement, but through:

- how parties perceive risk;
- how organisations respond to pressure;
- and how commercial decisions are strategically managed.

Even technically valid claims may fail because of:

- weak communication;
- delayed action;
- or poor strategic framing.

Conversely, many disputes are resolved not through strict legal entitlement, but through:

- commercial pragmatism;
- risk balancing;
- and negotiated compromise.

10.7 Data and Documentation Will Increasingly Define Claim Outcomes

As projects become more complex, reliable data increasingly determines:

- claim credibility;
- causation strength;
- and dispute defensibility.

Historically, many claims relied heavily upon:

- narrative explanation;
- reconstructed chronology;
- and subjective interpretation.

Future project environments will depend more heavily upon:

- integrated operational records;
- structured productivity analysis;
- digital project systems;
- and real-time data visibility.

This transformation carries significant implications.

The future strength of construction claims will depend not merely upon:

- legal drafting;

but also upon:

- quality of operational information;
- reliability of project systems;
- and organisational capability to interpret data effectively.

10.8 Practical Insight — Decision Timing

This issue can be understood through a simple practical observation.

At a traffic signal, movement occurs through clearly defined phases:

- stop;
- prepare;
- and proceed.

Ignoring these signals creates:

- confusion;
- inefficiency;
- and potential conflict.

Construction claims often evolve similarly.

Projects involve multiple stages requiring:

- timely notice;
- early recognition;
- structured evaluation;
- and appropriate decision-making.

For example:

- at notice stage → action must be initiated;
- at impact stage → operational implications must be assessed;
- at evaluation stage → entitlement and strategy must be determined.

When decisions are delayed:

- records deteriorate;
- causation weakens;
- productivity loss accumulates;
- and disputes become substantially more difficult to resolve.

Timeliness itself becomes a major component of effective project administration.

Key Insight

Timely decisions are often as important as technically correct decisions.

10.9 Construction Claims as Reflections of Project Management Quality

A broader observation emerging throughout this book is that:

construction claims frequently reflect the overall quality of project management itself.

Projects with:

- clear communication;
- structured documentation;
- timely decision-making;
- integrated commercial awareness;
- and collaborative problem-solving;

usually experience:

- lower dispute escalation;
- stronger claim defensibility;
- and more commercially balanced outcomes.

Conversely, projects characterised by:

- fragmented communication;
- reactive management;
- delayed approvals;
- defensive administration;
- and weak operational visibility;

frequently experience:

- cumulative disruption;
- prolonged conflict;
- and weakened commercial performance.

Claims are not merely legal events occurring after project problems emerge.

They are often symptoms of:

- broader project-management behaviour;
- organisational culture;
- and operational coordination.

10.10 Final Reflection

Construction projects involve:

- uncertainty;
- interdependence;
- human judgement;
- evolving constraints;
- and continuous adaptation.

No contract, however detailed, can fully eliminate these realities.

Accordingly, successful claim management depends not only upon:

- contractual rights;
- or legal interpretation.

It depends equally upon:

- operational awareness;
- behavioural discipline;
- structured documentation;
- commercial judgement;
- and timely strategic action.

The deepest lesson emerging from practical construction claims is therefore this:

Construction claims are not merely about proving entitlement after disruption occurs.

They are fundamentally about how projects recognise, manage, communicate, and respond to changing realities throughout execution.

Final Conclusion

Construction claims occupy a unique position within project management because they sit at the intersection of:

- engineering;
- law;
- commercial management;
- operational execution;
- and organisational behaviour.

Throughout this book, it has been demonstrated that:

construction claims cannot be properly understood through contractual interpretation alone.

While contracts establish:

- rights;
- obligations;
- procedures;
- and risk allocation frameworks;

actual claim outcomes are shaped equally by:

- execution realities;
- productivity behaviour;
- documentation quality;
- financial pressure;
- and human decision-making.

The case studies and analytical discussions throughout this work reveal several consistent realities.

First, disruption rarely arises from a single isolated event.

More commonly, it evolves through:

- cumulative operational interference;
- fragmented sequencing;
- changing site conditions;
- environmental constraints;
- delayed approvals;
- or continuous productivity deterioration.

Second, many claims fail not because entitlement is absent, but because:

- notices are delayed;
- causation is weak;
- records are incomplete;
- or commercial strategy is ineffective.

Third, construction projects repeatedly demonstrate that:

practical execution realities often extend beyond simplistic contractual assumptions.

This becomes particularly evident in disputes involving:

- incomplete information;
- safety-related restrictions;

- operational feasibility;
- and productivity loss.

The analysis further highlights that:

project value extends beyond direct cost.

Time savings, continuity of execution, risk mitigation, and avoidance of operational deterioration, often carry significant commercial importance, even where they are not immediately visible within conventional valuation frameworks.

An equally important conclusion emerging throughout this work is that:

construction claims are deeply behavioural.

The manner in which:

- project teams communicate;
- management responds to emerging disruption;
- documentation is maintained;
- and negotiations are approached;

frequently determines practical claim outcomes more than strict contractual wording alone.

Looking forward, the future of construction claims will depend increasingly upon:

- structured digital systems;
- integrated project data;
- proactive risk monitoring;
- and collaborative project administration.

However, despite technological advancement, one reality will remain unchanged:

successful claim management will always require sound judgement.

Ultimately, construction claims are not merely mechanisms for recovery after problems occur.

They are reflections of how effectively projects:

- recognise uncertainty;
- manage operational change;
- preserve commercial balance;
- and align contractual intent with execution reality.

Professional Credit References

This book is primarily based on practical project experience, professional observations, and case-based analysis developed over several decades of involvement in construction commercial management, contract administration, claims management, and dispute resolution.

In addition to practical experience, the following professional publications, protocols, and industry resources have contributed to the broader conceptual understanding of construction claims, disruption analysis, contractual risk allocation, and commercial management principles discussed in this book.

Industry Protocols and Standard References

- Society of Construction Law (SCL) — Delay and Disruption Protocol
- AACE International Recommended Practices
- FIDIC Conditions of Contract
- NEC Engineering and Construction Contract
- Standard industry practices relating to Extension of Time (EOT), disruption analysis, and productivity assessment

Professional Publications and Reference Works

- Pickavance, Keith — *Delay and Disruption in Construction Contracts*
- Haidar, Ali D. and Barnes, Peter — *Delay and Disruption Claims in Construction*
- Davison, R. Peter — *Evaluating Contract Claims*

Broader Professional Influence

The broader reflections within this book have also been influenced by:

- practical construction project experience;
- interaction with contractors, consultants, employers, and dispute professionals;
- industry practices across infrastructure and building projects;
- and evolving approaches to commercial and contractual management within the construction industry.

Appendix

Practical Reference Notes for Construction Claims

Appendix A — Typical Indicators of Emerging Disruption

The following operational signs frequently indicate emerging disruption within projects:

Operational Indicators

- declining labour productivity;
- repeated work stoppages;
- fragmented work fronts;
- increased idle time;
- repeated resequencing;
- excessive rehandling of material;
- abnormal equipment standby.

Administrative Indicators

- delayed approvals;
- accumulation of pending RFIs;
- inconsistent instructions;
- repeated design clarification;
- delayed decision-making.

Commercial Indicators

- delayed subcontractor payment;
- labour instability;
- vendor reluctance;
- irregular supply-chain behaviour;
- increased recovery negotiation.

Appendix B — Typical Records Useful for Disruption Claims

Effective disruption substantiation frequently depends upon:

- daily progress reports;
- labour deployment records;
- equipment utilisation logs;
- progress photographs;
- productivity tracking;
- weather records;
- access restriction records;
- site instruction registers;
- programme updates;
- and correspondence chronology.

Appendix C — Typical Causes of Weak Claims

Claims frequently weaken because of:

- delayed notice;
- absence of contemporaneous records;
- unclear causation;
- mixing of claim categories;
- poor quantification;
- exaggerated valuation;
- and inconsistent narrative.

Appendix D — Practical Questions During Claim Evaluation

During practical claim assessment, the following questions are often critical:

1. What specifically changed from the original execution plan?
2. Was the impact foreseeable at tender stage?
3. Could the Contractor reasonably mitigate the impact?
4. Was productivity measurably affected?
5. Is the causal linkage clearly demonstrable?
6. Were notices and records maintained contemporaneously?
7. Does the valuation logically arise from the demonstrated impact?

Appendix E — Consolidated Practical Insights

Practical Insight	Theme	Related Discussion
Milk Heating	Early warning and intervention	Emerging disruption
Drizzle vs Downpour	Duration vs intensity	Weather disruption
Traffic Signal	Timely decisions	EOT and claim management
Butter Clarification	Knowing when to conclude	Claim preparation

Appendix F — Final Practical Reflection

Construction projects are ultimately executed not in ideal theoretical environments, but within:

- changing conditions;
- operational pressure;
- commercial uncertainty;
- and human limitations.

Accordingly, effective claim management requires more than:

- contractual knowledge;
- or technical expertise.

It requires:

- practical judgement;
- operational awareness;
- behavioural discipline;
- and the ability to recognise how seemingly small events progressively evolve into major commercial consequences.

The most successful projects are therefore not necessarily those that avoid all disruption.

Rather, they are those capable of:

- recognising emerging issues early;
- responding proportionately;
- preserving evidence systematically;
- and maintaining commercial balance throughout execution.

ABOUT THE AUTHOR

SHREEDHAR BHAT is a construction professional with nearly Four decades of experience across:

- project management;
- commercial management;
- contract administration;
- claims management;
- valuation;
- and dispute resolution.

His professional exposure spans:

- residential;
- commercial;
- industrial;
- and large-scale infrastructure projects.

He is:

- a Fellow of the Royal Institution of Chartered Surveyors (RICS);
- a Fellow of the Institute of Construction Claims Practitioners;
- a Fellow of The Indian Institution of Valuers;
- a Qualified to be an Independent Director (IICA);
- an IIAM/APCAM Accredited Arbitrator;
- a Chartered Valuer;
- and a Certified professional in Alternative Dispute Resolution and Enterprise Risk Management.
- Corporate Trainer

Over the course of his professional career, he has been closely involved in:

- construction claims;
- disruption analysis;
- contract strategy;
- risk allocation;
- and commercial management under complex project environments.

His work combines:

- technical understanding;
- contractual interpretation;
- commercial pragmatism;
- and practical project execution insight.

This manuscript reflects his broader professional interest in understanding how:

- contractual frameworks;
- execution realities;
- organisational behaviour;
- and commercial decision-making

interact in shaping construction claim outcomes.

Made in the USA
Monee, IL
20 September 2026

63298132R00075